颂扬多元化——当今学校建筑

In Praise of Diversity-School Architecture Today

[英]安娜·鲁斯等 | 编

曹麟　吴美萱 | 译

大连理工大学出版社

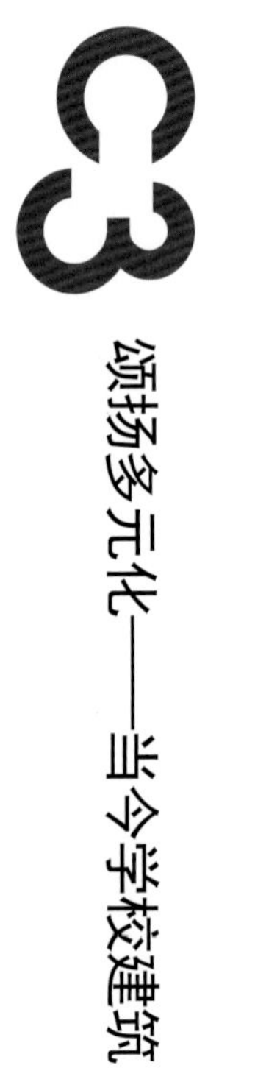

004 社会建筑——行业的觉醒 _ Michèle Woodger

哈德逊园区的Vessel和Shed大楼

010 向公众开放的哈德逊园区的Vessel和Shed大楼_YuMi Hyun

014 Vessel大楼_Heatherwick Studio

030 The Shed大楼_Diller Scofidio + Renfro

公共图书馆的社会资本建设

044 公共图书馆的社会资本建设_Davide Pisu

050 Oodi中央图书馆_ALA Architects

076 卡尔加里新中央图书馆_Snøhetta + DIALOG

回念：建筑嵌入、连续性和变化

098 回念：建筑嵌入、连续性和变化_Gihan Karunaratne

104 大馆传统与文化中心_Herzog & de Meuron

122 格拉纳达费德里科·加西亚·洛尔卡中心_MX_SI (Mendoza Partida + BAX studio)

138 卡托维兹西里西亚大学广播电视系_BAAS Arquitectura

颂扬多元化——当今学校建筑

150 颂扬多元化——当今学校建筑_Anna Roos

156 南港学校_JJW Arkitekter

172 圣伊西多尔学校集团扩建工程_ANMA

184 Žnjan-Pazdigrad小学 - Pazdigrad Primary School_x3m

202 TTC槟榔精英幼儿园_KIENTRUC O

216 Nová Ruda幼儿园_Petr Stolín Architekt

230 建筑师索引

In Praise of Diversity – School Architecture Today

004 Social architecture – an industry awakening _ Michèle Woodger

Vessel and The Shed at Hudson Yards

010 Vessel and The Shed at Hudson Yards open to the public _ YuMi Hyun

014 Vessel _ Heatherwick Studio

030 The Shed _ Diller Scofidio + Renfro

Building the Social Capital in the Public Libraries

044 Building the Social Capital in the Public Libraries _ Davide Pisu

050 Oodi Central Library _ ALA Architects

076 Calgary's New Central Library _ Snøhetta + DIALOG

Palimpsest: Architectural Insertions, Continuity and Change

098 Palimpsest: Architectural Insertions, Continuity and Change _ Gihan Karunaratne

104 Tai Kwun, Center for Heritage & Arts _ Herzog & de Meuron

122 Federico García Lorca Center in Granada _ MX_SI (Mendoza Partida + BAX studio)

138 Silesia University's Radio and TV Department in Katowice _ BAAS Arquitectura

In Praise of Diversity – School Architecture Today

150 In Praise of Diversity - School Architecture Today _ Anna Roos

156 South Harbor School _ JJW Arkitekter

172 Saint Isidore School Group Extension _ ANMA

184 Žnjan - Pazdigrad Primary School _ x3m

202 TTC Elite Ben Tre Kindergarten _ KIENTRUC O

216 Kindergarten Nová Ruda _ Petr Stolín Architekt

230 Index

社会建筑——行业的觉醒
Social architecture – an industry awakening

Michèle Woodger

"建筑不仅仅是一种用于表现的雕塑。它们使存在于社会中的个人和集体意愿变得显而易见。而且伟大的建筑还能赋予我们以希望。伟大的建筑可以治愈我们的伤痛。" 这段肺腑之言是在2016年，由总部位于美国和卢旺达的MASS设计集团（一家国际知名的、非营利性并专注于社会项目的公司）的创始人迈克尔·墨菲发出的。这些建筑包括卢旺达的布塔罗地区医院、海地的GHESKIO结核病医院和建在马拉维的孕妇候诊村。[1]

MASS是Model of Architecture Serving Society的字母缩写，意思是以建筑设计来服务社会的模式。它是过去十年中出现的几种建筑实践机构之一，将社会参与放在首位。正如墨菲在TED会议的其他讨论中所发表的热情演讲所说的："为什么最好的建筑师、最伟大的建筑……是如此罕见，而且似乎服务于如此少的人群？拥有了这些具有创造性的人才，我们还能做些什么呢？"[2]

墨菲的观点并不是孤立的。近年来，越来越多的建筑从业者公开拒绝了"明星建筑师"这种有排他性的志向，他们开始赞同一种新的信念，即建筑的"伟大"是通过回馈社会来实现的。

建筑已经发生了更具社会意识的转变，这也许并不奇怪。在过去的20年中，出现了巨大的技术变化（互联网、社交媒体、人工智能）和社会动荡（"9·11事件"后的世界、2008年的经济衰退、"阿拉伯之春"运动等）。出生于

"Buildings are not simply expressive sculptures. They make visible our personal and our collective aspirations as a society. Great architecture can give us hope. Great architecture can heal." Spoken in 2016, these are the heartfelt words of Michael Murphy, founding director of US and Rwanda-based MASS Design Group, an internationally renowned non-profit practice focusing on socially-oriented projects. These include the Butaro District Hospital in Rwanda, the GHESKIO Tuberculosis Hospital in Haiti, and the Maternity Waiting Village in Malawi. [1]

MASS, which stands for Model of Architecture Serving Society, is one of several architecture practices to emerge in the last decade which place social engagement at its heart. As Murphy warmly says elsewhere in the same TED discussion: "Why is it that the best architects, the greatest architecture…is also so rare, and seems to serve so very few? With all of this creative talent, what more could we do?"[2]

Murphy is not alone in his viewpoint. In recent years an increasing number of practitioners are overtly rejecting the exclusive aspirations of "starchitects", in favour of the belief that architectural "greatness" is achieved through what it gives back to society.

It is perhaps not surprising that architecture has taken a more socially conscious turn. The last twenty years have seen enormous technological changes (the Internet, social media, AI) and social upheavals (the post 9/11-world, the 2008 economic downturn, the Arab Spring, etc.). Those who were born between 1980 and 2000 – today's early-mid career professionals – have come of age in this milieu. And whilst it is impossible to generalise, it is often said that this millennial generation is more entrepreneurial, more awake to social concerns, and understandably more suspicious (even cynical) of

1980年到2000年之间的这群人——今天的中早期职场专业人士——都是在这种大环境中长大的。

虽然不可能一概而论，但人们常说，千禧年的一代人更具创业精神，对社会问题更为清醒，对已建立的自上而下的等级制度持有更加怀疑（甚至是愤世嫉俗）的态度，这或许就是为什么某些非正统和非正式的商业模式如今都形成了一种常态：例如，共享经济、公民新闻和众筹。据《福布斯》杂志报道，千禧一代"更愿意与具有亲社会信息、可持续的制造方法和道德商业标准的企业和品牌开展业务合作"。[3] 千禧年涌现的第一批建筑师现在已经具备了一定的专业资格，积累了丰富的经验，并进入了可以把自己的理想和创造性思维付诸实践的职业生涯阶段。

这并不是某些特别热心公益的组织的孤立行动。富有社会责任感的建筑形式现已成为教育和实践的关键组成部分；2018年，RIBA将这一点作为其继续职业发展计划的一个强制性要素，并指出："为社会服务的建筑是为了了解建筑为个人和社区所带来的社会价值、经济效益以及环境效益的。[建筑应该]改善人们的生活，增强社会认同感，促进和培养凝聚力和幸福感。"同时，建筑师应该是"一个榜样，既要以身作则，又要利用他们在项目中的角色作用，创造积极影响其他人的机会"。[4]

该行业的社会意识开始觉醒的迹象已经出现了差不多十年。2011年，纽约现代艺术博物馆的展览"小规模、大

established top-down hierarchies, which is maybe why certain unorthodox and informal business models are now the norm: the sharing-economy, citizen journalism and crowd-funding for instance. According to Forbes, millennials "prefer to do business with corporations and brands with pro-social messages, sustainable manufacturing methods and ethical business standards". [3] The first wave of architect millennials have now qualified, accrued sufficient experience, and reached a career stage where they can put their ideals and out-of-the box thinking into practice.

It is not a case of certain particularly public-spirited organisations acting in isolation. Socially responsible architecture has now become a key component of education and practice; in 2018 the RIBA introduced this as a mandatory element of its continuing professional development program, stating: "Architecture for social purpose is about understanding the social value and economic and environmental benefits that architecture brings for individuals and communities. [Architecture should] improve people's lives, enhance social identity, enable and foster cohesion and wellbeing". Meanwhile architects should be "role models, both leading by example and using their position on a project as an opportunity to positively influence other[s]." [4]

Signs that the industry was beginning to awaken its social consciousness have been occurring for about ten years. In 2011, MoMA's exhibition 'Small Scale, Big Change: New Architectures of Social Engagement' presented eleven projects from around the world, such as primary schools in Burkina Faso and Bangladesh and housing projects in Chile and Lebanon. The projects were chosen for their commitment to social responsibility. According to the curators Andres Lepik and Margot Weller: "the

变革：社会参与的新架构"展示了来自世界各地的11个项目，例如，布基纳法索和孟加拉国的小学，以及智利和黎巴嫩的住房项目。选择这些项目是因为它们对承担社会责任所做出的承诺。馆长安德烈斯·勒皮克和玛戈特·韦勒说："这里提到的建筑师对宏大的宣言或乌托邦式的理论不感兴趣。相反，他们对激进实用主义的承诺可以在他们所实现的项目中看到，这些项目为已知的需求提供了切实可行的解决方案，同时也旨在对他们工作的社区产生更广泛的影响。"[5]

这其中一个典型的例子是由弗雷德里克·德鲁特、安妮·拉卡顿和让·菲利普·瓦萨尔在2006年至2011年对巴黎一座始建于20世纪60年代的16层住宅楼Tour Bois le Prétre进行的改造工程。它被公认为是社会住房改造的成功案例。原先的大楼绰号叫作"恶魔岛"，建筑师采用大窗户、玻璃阳台、扩展地板层对大楼进行了更新，引入更多的自然光线和通风，还能全方位欣赏到巴黎市区的美景。预制构件的使用大大减少了成本的支出，让居民们得以继续在此居住，而最初的提议是对其进行拆除。引用德鲁特的话说："居民们觉得他们的想法得到了倾听、认可和考虑。[这种方法可以]在世界任何地方使用。作为建筑师，要慷慨大度，在追求良好的经济效益的同时也要尽可能多地付出。"随后，Tour Bois de Prêtre的模式已应用于从南美到法国南部的很多项目。[6]

architects highlighted here are not interested in grand manifestos or utopian theories. Instead, their commitment to a radical pragmatism can be seen in the projects they have realized [which] offer practical solutions to known needs, but also aim to have a broader effect on the communities in which they work."[5]
One featured example is the 2006-2011 transformation of Tour Bois le Prêtre – a 16-story 1960s housing block in Paris – by Frédéric Druot, Anne Lacaton, and Jean Philippe Vassal. It is widely considered a success story of social housing renovation. The original tower, nicknamed "Alcatraz", was updated with large windows, glazed balconies, extended floor-plates, more natural light and ventilation and 360-degree views across Paris. The use of pre-fabricated components greatly reduced the expenditure, allowing residents to remain in their homes, whereas the original proposal had been demolition. Druot is quoted as saying: "The residents felt that they were listened to, recognised and considered. [This approach can be used] anywhere in the world. It's a matter of …being generous as an architect, to give as much as possible in the pursuit of good economics." Subsequently the model for Tour Bois de Prêtre has been applied to the projects from South America to southern France.[6]
Another defining moment for the industry was the surprising decision to award the 2015 Turner Prize (usually reserved for individual artists) to the architecture collective Assemble, known for their collaborative regeneration project in Granby Four Streets, Liverpool.[7] This project was the result of a two-decade long battle by the inhabitants of a neglected and derelict area. The once tight-knit and ethnically diverse community found themselves scattered across the city as homes were vacated and

蒙罗伊别墅，增量住宅项目，包含93个住宅单元，ELEMENTAL设计，智利伊基克，2004年
Quinta Monroy, an incremental housing project of 93 units by ELEMENTAL, Iquique, Chile, 2004

用户干预改造之后的蒙罗伊别墅
Quinta Monroy after the users' intervention

该行业的另一个关键时刻是将2015年特纳奖（通常为个人艺术家保留）授予合作完成利物浦的格兰比·福尔街重建项目的建筑团体Assemble，这个决定有点令人惊讶。[7]该项目的位置是在一个被忽视和遗弃的地区里，长居于此的居民们为此进行了长达20年的斗争。这个曾经联系紧密、种族多样的社区发现他们自己分散在城市各处，因为这里的房屋被腾空，部分建筑被拆除。剩下的居民组成了一个联盟，并致力于清理和耕植该地区。Assemble利用这一有机品质和足智多谋的精神，目的是充满敏感度地翻新10个住宅、重燃公共空间的活力，并为当地人提供培训和机会。

2016年，普利兹克奖被授予智利建筑师亚历杭德罗·阿拉维纳。评委们认为他"代表了一种更具社会责任感的建筑师的复兴，尤其是他长期致力于解决全球住房危机，为所有人创造更好的城市环境"。特别值得肯定的是，专家组称赞了他的建筑工作室ELEMENTAL所做的工作，他把工作室描述为"操作团"（与"智囊团"恰恰相对，再次重申向务实的建筑设计的转变）。ELEMENTAL专注于与社会相关的项目；例如，创新的"增量住宅"，这种设计为居民提供了空间，使得居民自己可以完成家园的设计，从而感受到一种成就感和投资感。据评委会称："建筑师的角色现在正受到挑战，以满足更大的社会和人道主义需求，而亚历杭德罗·阿拉维纳对这一挑战有着清楚的认识，并慷慨、充分地回应了这一挑战。"[8]

partially demolished. The remaining residents formed an association and dedicatedly cleared and planted-up the area. Assemble harnessed this organic quality and resourceful spirit, in order to sensitively renovate ten homes, revitalise public spaces, and provide training and opportunities for locals.
In 2016 the Pritzker Prize was awarded to Chilean architect Alejandro Aravena. The judges felt that he "epitomizes the revival of a more socially engaged architect, especially in his long-term commitment to tackling the global housing crisis and fighting for a better urban environment for all." In particular, the panel praised the work of his practice ELEMENTAL, which he describes as a "do-tank" (as opposed to a "think-tank", again reiterating the shift towards a pragmatic architecture). ELEMENTAL focuses on socially relevant projects; such as the innovative "incremental housing" whereby designs leave space for inhabitants to complete the homes themselves thereby feeling a sense of achievement and investment. According to the jury: "The role of the architect is now being challenged to serve greater social and humanitarian needs, and Alejandro Aravena has clearly, generously and fully responded to this challenge."[8]
There are several key approaches which apply, to a greater or lesser degree, to almost all projects falling under the category of 'social architecture': the desire to improve the living or working conditions for marginalised social groups; the involvement of, or collaboration with, the community; a focus on localism, be it through sourcing materials or drawing inspiration from vernacular building typologies; and the belief that the innovation – developed and applied on a small scale – could in theory be adopted globally, thus instigating a wave of positive change around the world.

有几种关键的方法，或多或少地适用于几乎所有属于“社会建筑”范畴的项目：希望改善边缘化社会群体的生活或工作条件；参与或与社区合作；注重地方主义，无论是通过采购材料还是从当地建筑类型中汲取灵感；以及相信创新——小规模的开发和应用——理论上可以在全球范围内被采用，从而在世界范围内掀起一波积极变革的浪潮。

到目前为止，情况还不错。上面给出的建筑实例都是令人鼓舞的。但从学术方面来看，“社会建筑”确实有一些贬低者。2017年，里克特、戈贝尔和格鲁巴尔在《城市》杂志上撰文指出了几个问题（尽管没有挑出任何一个特定的项目或工作室进行批评）。

在这种大环境之下，“全球化”就是一种自相矛盾的概念：一方面，它被描绘成一种具有威胁的、同质化的团体力量；另一方面，它作为项目的最终目标，一个微小的动作所引起的涟漪也会波及全球。对地方主义的热爱并不能轻松应对这些全球化的志向。作者指出，由于社会建筑项目通常是“具有更大反响的小型活动……小规模的参考也提供了一种不必与更广泛的政治、文化和经济大环境（超出项目的范围）接触的便捷方式。”在某些情况下，个人项目是回避从源头上解决更大社会问题的变通方法，而不是起到“社交孤岛”的作用。[9]此外，特定项目的直接成功描

So far so good, and the examples given above are nothing short of inspirational. But as an academic concern, "Social Architecture" does have some detractors. Writing in the journal *City*, in 2017, Richter, Göbel and Grubbauer identify several problematic issues (albeit without singling out any one particular project or practice for criticism).

One paradoxical concept is that of "globalism" within this context: on the one hand it is pictured as a threatening, homogenising, corporate force; on the other, as the ultimate goal of the outreach of the projects, whereby ripples of small actions take on worldwide proportions. A love of localism does not sit easily with these global ambitions. The authors point out that since social architecture projects are "often small activities with projected larger reverberations…the affirmative reference to small scales also offers a convenient way of not having to engage with the broader political, cultural and economic context (beyond a specific project) in substantial ways." In some cases, individual projects are workarounds that shy away from addressing bigger social issues at source, instead acting as "islands of social engagement". [9] Furthermore, the immediate successes of given projects paint a misleadingly positive picture; many are still in their infancy to be evaluated for a long term.

In the same volume, Gribat and Meireis argue that there is insufficiently rigorous treatment of key architectural theories within the debate. Aesthetics is one example: "It is taken for granted that well-designed, beautiful buildings "do good" – without questioning what appears as well-designed to whom, under which conditions and in which context." [10]

Most troublingly they posit the risk of a power imbalance – with paternalistic, post-colonial implications – when Western architects undertake projects in developing nations: "In the social architecture

绘了一幅有误导性的积极画面；许多项目仍处于不成熟期，有待长期评估。

在同一本书中，格里巴特和梅伊雷斯认为在辩论中对关键建筑理论的处理不够严谨。美学就是一个例子："毫无疑问，经过精心设计的美观建筑产生良好的效果是理所当然的——无论项目的设计初衷、项目条件和项目环境如何。"[10]

最令人不安的是，当西方建筑师在发展中国家做项目之时，他们假定存在权力失衡的风险——带有家长式作风的、后殖民主义的意味："在关于社会建筑的讨论中，即使有人提出要求，希望建筑师和设计师的设计多亲近他们所服务的对象，建筑师和设计师仍然处于一个比较强势的位置。"[10]即使建筑师推动并且采用了当地传统的、扎根社区的设计方法，最后仍然是他们的设计和专业知识为他们提供了解决方案，挽救了局面。

作者认为，将"社会建筑"作为学术话题是一种进步，"以一种更具反思性的方式进一步发展，例如，放弃对项目的限制，也不要限制建筑师改变世界的力量。"[10]确实，乐观主义和理想主义是在建筑和社会中都需要培养和鼓励的品质，但重要的是，这些行为不能越界，变得幼稚，或者更糟——变得傲慢。有鉴于此，让我们希望无论是在社会事务方面，还是在自我意识方面，这个行业都能继续觉醒。

debate, architects and designers remain in a powerful position, even if they are asked to move a bit closer to the people they serve".[10] Even as architects promote and cultivate local, traditional, community-based approaches, it is still their designs and expertise which ultimately provide the solution that saves the day.

A way forward, the authors argue, is for "Social Architecture" as an academic discourse "to be developed further in a more reflective manner, for instance, in giving up the fixations on projects and on the power of architects to change the world."[10] Indeed, while optimism and idealism are qualities to be nurtured and encouraged within architecture – and in society – it is important that these should not cross the boundary into naivety, or worse, arrogance. With this in mind, let us hope that the industry continues its awakening, both in social matters, and in self-awareness.

1. https://www.ted.com/talks/michael_murphy_architecture_that_s_built_to_heal/transcript?language=en#t-126846
2. https://www.ted.com/talks/michael_murphy_architecture_that_s_built_to_heal?language=en
3. https://www.forbes.com/sites/sarahlandrum/2017/03/17/millennials-driving-brands-to-practice-socially-responsible-marketing/#1f5242464990
4. https://www.architecture.com/knowledge-and-resources/knowledge-landing-page/what-is-architecture-for-social-purpose
5. https://www.moma.org/calendar/exhibitions/1061
6. https://www.rightanglestudio.com.au/urbanresearchjournal/field-notes/architecture-of-opportunity-tour-bois-le-pretre/
7. https://www.tate.org.uk/whats-on/other-venue/exhibition/turner-prize-2015/turner-prize-2015-artists-assemble
8. https://www.pritzkerprize.com/laureates/2016
9. Anna Richter, Hanna Katharina Göbel & Monika Grubbauer, 'Designed to improve? The makings, politics and aesthetics of 'social' architecture and design,' City, 02 November 2017, Vol.21(6), pp.769-778
10. Gribat, N. and Meireis, S., 'A critique of the new 'social architecture' debate: Moving beyond localism, developmentalism and aesthetics', City, 02 November 2017, Vol.21(6), pp.779-788

向公众开放的哈德逊园区的 Vessel 和 Shed 大楼

Vessel and The at Hudson Ya open to the pu

沿着哈德逊河，临近高线公园的北端，杂草丛生的一片荒地已经变成纽约最热门的新地段。

美国最大的私人开发项目——哈德逊园区投资约250亿美元。哈德逊园区拥有超过167万平方米的混合用途开发面积，包括100多家商店、一系列餐厅、近4000套住宅、56600m²的公共开放空间、豪华酒店、医疗设施、体育中心、表演艺术场馆和一所可容纳750个学位的公立学校。

3月15日，第11大道以东的哈德逊园区一期工程在一个盛大的开幕式上向公众展示。开放区域包括零售和餐厅设施、景观广场和花园，以及坐落于20000m²公共广场和通往东部园区的花园中心的"Vessel"项目。继Vessel项目之后，巨大的新艺术中心，即紧挨着高线公园旁边的东部园区的The Shed，于4月5日开门迎客。

到2025年底，当西部园区及16栋建筑全部建成后，预计每天都会有超过12万名游客、美食家、时尚爱好者和公园爱好者参观哈德逊园区。

Along the Hudson River, near the northern end of the High Line park, overgrown and neglected land has been transformed into the hottest new place in New York.

Around $25 billion was invested in the Hudson Yards, the largest private development in the USA. With more than 1,670,000m² of mixed-use development, Hudson Yards includes over 100 shops, a collection of restaurants, approximately 4,000 residences, 56,600m² of public open space, luxury hotels, medical facilities, offices, performing arts venues and a 750-seat public school.

Hudson Yards' phase one, east of 11th Avenue, was unveiled to the public at a grand opening on March 15th. Opening areas include the retail and restaurant provision, the landscaped plaza and gardens, and "Vessel" that sits at the center of a 20,000m² Public Square and Gardens to the Eastern Yard. Following Vessel, the massive new art center, The Shed, which borders the High Line along the Eastern Yard, opened its doors on April 5th.

By the end of 2025, when the development of the Western Yard and all 16 buildings are completed, more than 120,000 tourists, gourmets, fashion lovers, and park enthusiasts will be expected to visit Hudson Yards every day.

Vessel, The Climbable Sculpture

Vessel, designed by British architect Thomas Heatherwick and his studio, was finally completed after four years of

Shed
ds
lic

©Related Oxford

©Related Oxford

Vessel——可攀爬的雕塑

这座由英国建筑师托马斯·赫斯维克和他的工作室设计的Vessel，经过四年的建造，终于完工了。受印度阶梯井的启发，这座建筑由一个钢框架制成，框架外部覆盖着镀铜的钢质覆层。它由一个直径为15m的底座处开始上升，顶部扩大到45m。这座标志性建筑物由154段错综复杂的楼梯相互连接而成，几乎有2500个独立的台阶和80个楼梯平台，在公共广场和花园上方有近1.61km的垂直攀升高度，从而为人们提供了非同寻常的视野。

赫斯维克的设计从一开始就处于保密的状态。该项目的钢构件由意大利的专业制造商Cimolai生产，钢厂周围建有一道6m高的围栏，因此没有人能看到设计过程。这些钢结构的构件被运到曼哈顿的西区港口，在那里用驳船沿着哈德逊河用时5小时运输到达目的地。随着设计逐步被披露出来，围绕着设计开始出现了争议的声音。纽约市民和批评家称这个项目为“蜂巢”“烤肉串”，甚至是“垃圾桶”，此前就有消息称建造成本超过1.5亿美元——然而，回想当初，正如人们早期对巴黎埃菲尔铁塔的批评一样，人们对这个项目的看法应当也会有所改变。

construction. Inspired by Indian stepwells, Vessel is made of a steel frame covered with copper-clad steel claddings. It rises from a base that is 15m in diameter and widens to 45m at the top. Comprised of 154 intricately interconnecting flights of stairs – almost 2,500 individual steps and 80 landings – with nearly one mile of vertical climb above the public square and gardens, this landmark offers remarkable views.

Heatherwick's design was kept under wraps from the beginning. Vessel's steel components were produced by specialist fabricator Cimolai in Italy, and a 6-meter-high fence was built around the steelworks so that no one could see the design process. The parts of steel structure were brought to Manhattan's West Side harbor, where it was transported by barge for five hours along the Hudson River. As the design was gradually revealed, the controversy surrounding Vessel began to grow. New Yorkers and critics labelled Vessel a "beehive", a "kebab", and even a "garbage can", following news that the cost of construction was more than $150 million – however, in retrospect, just as with the early criticism of the Eiffel Tower in Paris, the perception of Vessel is expected to change.

"We've never designed anything like this before, and we'll probably never design anything like it again," said Heatherwick Studio's group leader and partner Stuart Wood. "People often ask us, what is this for? Is it a viewing platform? Where are you looking to? ... It's not a building, it's not a sculpture, it's not an artwork, and yet it has scale and relevance to all of those typologies. In a way, we're thinking of this as a piece of furniture. Its ongoing use will

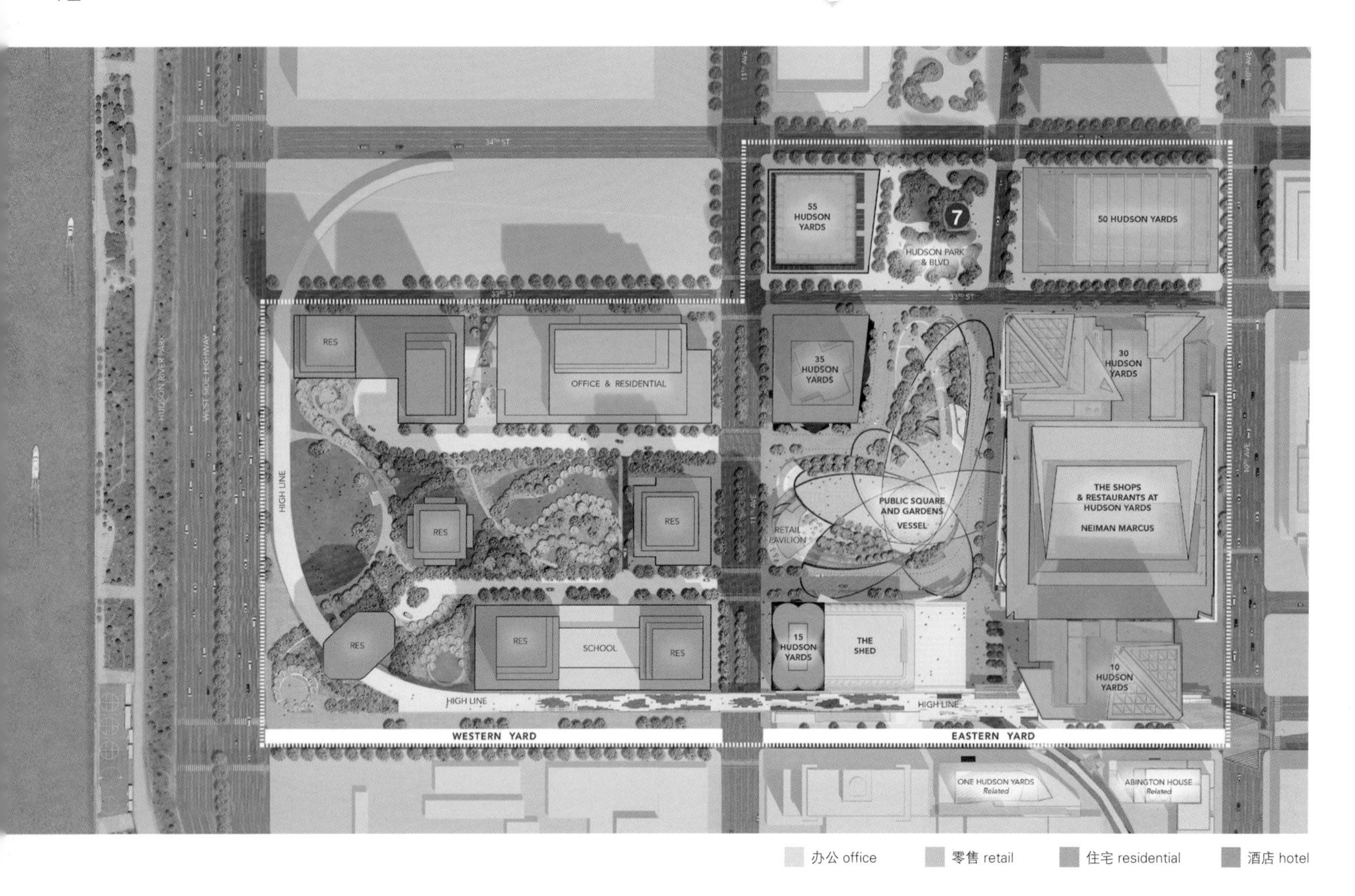

"我们以前从来没有设计过这样的作品，而且我们可能再也不会设计这样的作品了，"赫斯维克工作室的团队负责人兼合伙人斯图尔特·伍德说。"人们经常问我们，这是干什么用的？是观景台吗？你要观哪里的景？……它不是一座建筑，不是一座雕塑，也不是一件艺术品，但它与所有这些建筑类型都有着规模上的相似性和相关性。在某种程度上，我们认为这是一件家具。它的持续使用将产生自然演变的效果。"

在Vessel开放的那天，哈德逊园区的公共广场和花园里挤满了一大群人，他们毫不犹豫地在建筑物上爬上爬下。通过在线预订，Vessel可以免费预订游览，但票实在是太紧俏，在接下来的两周内就售罄了。

Shed，一个适应性强、可扩展的文化场所

该项目由Diller Scofdio + Renfro建筑事务所和罗克韦尔集团设计，是一个基于"移动结构"概念的新文化艺术空间。这个项目的灵感来自塞德里克·普莱斯的"游乐宫"，那是一座始建于20世纪60年代、坐落于伦敦的未建成的建筑，同时，该建筑的灵感还来自于高线公园和曼哈顿西区铁路站的工业历史。

evolve, quite naturally."

At the opening day of Vessel, the Public Square and Gardens of Hudson Yards were filled with a big crowd of people taking selfies without hesitation as they climb up and down the structure. Vessel can be booked for free with online reservations, but tickets are already popular enough to be sold out for the next two weeks.

The Shed, An Adaptable and Expandable Cultural Venue

Designed by Diller Scofidio + Renfro and Rockwell Group, The Shed is a new culture and art space based on the concept of a "moving structure". The Shed was inspired by Cedric Price's Fun Palace, an unrealized building-machine from 1960s London, as well as by the industrial past of the High Line and the West Side Railyard.

The building consists of a largely fixed main building and a giant retractable outer shell that can be rolled back for outdoor events. The shell, weighing more than 3,500 tons, has six 1.8m-wheels of 15 horsepower at the bottom. When these wheels move along the steel rails, the place where it was normally used as an outdoor plaza becomes

哈德逊园区鸟瞰图 Hudson Yards Aerial View, May 2018 (©Related Oxford)

这座建筑由一个很大程度上封闭的主楼和一个巨大的可伸缩外壳组成，这个外壳可以在举办户外活动时向后翻转。这个重达3500t的外壳底部由6个1.8m的轮子组成，可提供15马力的动力。当这些轮子沿着钢栏杆移动时，通常用作室外广场的地方就变成了室内空间。只需五分钟就可以将空间扩大一倍。项目的开放式基础设施可以在未知的未来保持永久存在性，并在规模、媒介、技术和艺术家不断变化的需求方面做出反应。

"在整个11年的建造过程中，这个项目一直在向公众敞开大门，这是一项持续不断的工作。" Diller Scofdio + Renfro建筑事务所的伊丽莎白·迪勒说，"我认为这座建筑是一个'基础设施的建筑'，它所展现的是肌肉，而不是脂肪，响应了艺术工作者进入我们无法预测的未来之后所提出的不断变化的需求。"

虽然Vessel这个项目的设计形式能给人带来强大的视觉刺激，但它所提供的系统和技术可以对不可预测的变化做出反应，而这种反应不仅仅是形式层面的。根据上述两位建筑师的话，Vessel可以说是一件"家具"，Shed可被称为是"基础设施的建筑"。两者孕育于同一地区，又都充满了独特的个性。它们在互相不可穿透的力量之间实现了一种平衡，已经成为哈德逊园区真正的标志性建筑。

an interior space. It takes only five minutes to double the space. The Shed's open infrastructure can be permanently flexible for an unknowable future and responsive to variability in scale, media, technology, and the evolving needs of artists.

"Eleven years in the making, The Shed is opening its doors to the public as a perpetual work-in-progress. I see the building as an 'architecture of infrastructure,' all muscle, no fat, and responsive to the ever-changing needs of artists into a future we cannot predict," said Elizabeth Diller of Diller Scofidio + Renfro.

While Vessel reveals formality as a powerful visual stimulant, The Shed provides systems and technologies that can respond to unpredictable changes rather than the form. According to the words of the two above-mentioned architects, Vessel is arguably "furniture" and The Shed claims to be "infrastructure". Both were born in the same place with totally distinct personalities. Balancing each other's impenetrable power, they have become new landmarks in Hudson Yards indeed. YuMi Hyun

Vessel 大楼

Vessel

Heatherwick Studio

当新曼哈顿遇见赫斯维克建筑事务所的Vessel攀登架

Vessel项目是一个16层的环形攀登架，这是一种新型的标志性公共建筑。它有2465个台阶，80个楼梯平台，可以俯瞰整个哈德逊河和曼哈顿。这是哈德逊园区开发项目的主要公共广场的重要特色，这个开发项目是美国历史上最大的房地产项目之一。该项目将曼哈顿+上西城以前的一个铁路站场改造成了一个全新的社区，拥有超过2.02ha的新公共场所空间和花园。赫斯维克建筑事务所应邀为哈德逊园区设计一个中央地标结构，用来欢迎到访游客进入该地区的中心区域，并在曼哈顿创造一个新的供大家使用的会面地点。设计挑战当中的一部分是要创造一些值得纪念的东西，而这些东西不会被周围高耸的楼群或火车站台上方新公共空间的规模所压制。为了探索不同的可能性，团队开始缩小参数：它应该是一个令人难忘的单一物体，而不是一系列分散在整个空间的物体；也不应该是一个具有惰性的静态雕塑，它应该是一个鼓励活动和参与的社交所在——它应该具有趣味性。

让我们看看城市里人们自然聚集的地方，底层的基础设施通常很简单——比如拿楼梯来说，看看位于罗马的著名的"西班牙台阶"。通过进一步研究这一类型的设计，工作室设计出了具有传统概念的印度阶梯井；这些台阶具有一个错综复杂的石阶网络，以便当水库的水位发生变化的时候，表面仍然可以作为输水通道来使用。然而，如同圆形剧场一样，阶梯井的焦点在于它的中心，工作室想要创造一种既外向又内向的体验。

通过打开台阶之间的空隙来创建一种三维格架，公共广场就可以向上延伸，创造出长度超过1.6km的路径，并可以以不同的方式进行探索。为了通过154个相连的台阶来创建阶梯井的连续几何图形，建筑的结构必须是自支撑式的——这就需要谨慎的结构解决方案，不能借助额外的柱和梁。这是通过在每对楼梯之间插入一根钢脊梁来解决的，这样就在"向上"和"向下"之间创造出了自然的划分。这种结构的原始焊接钢架被暴露在外，从而使建筑产生清晰感和完整感。楼梯的底部用深铜色金属包裹，使其与周围的建筑区分开来。

从建筑的接缝到扶手，Vessel的每一个部件都是定制的。75个巨大的钢构件由专业制造商Cimolai公司在威尼斯制造，然后从意大利分6批运来，再用驳船经过哈德逊河运抵现场。现场的组装过程耗时3年。然而，尽管项目的体积很大，但它的设计还是按人体的尺度进行的，以便于供纽约当地人和游客攀爬、探索和欣赏风景。虽然这是一种简单的结构，但人群和底部的广场让它充满了活力。

New Manhattan encounters on Heatherwick's Vessel climbing frame

Vessel is a new type of public landmark – a 16-storey circular climbing frame, with 2465 steps, 80 landings and views across the Hudson River and Manhattan. It is the central feature of the main public square in the Hudson Yards development, one of the largest real estate projects in American history, which is transforming a former rail yard in Manhattan's Upper West Side into a completely new neighborhood, with more than five acres of new public spaces and gardens. Heatherwick Studio was asked to design a centerpiece for Hudson Yards, something that would welcome visitors into the heart of the district and create a new place to meet in Manhattan. Part of the challenge was to create something memorable that would not be overwhelmed by the surrounding cluster of towers, or the scale of the new public space above the train platform. Exploring different possibilities, the team started to narrow the parameters: it

SEPHORA
SEPHORA
12
11
10
9
8
7
6
5
4
3
2
1

should be a memorable single object, not a series of objects dispersing throughout the space; rather than an inert, static sculpture, it should be a social encounter, which encourages activity and participation – it should be fun.
Looking at the places in cities where people naturally congregate, the underlying infrastructure is often simple – a staircase, for example, such as the famous Spanish Steps in Rome. Researching this typology further, the studio explored traditional Indian stepwells; they have an intricate network of stone stairs, so that as the water level in the reservoir changes, the surface is still accessible. However, like an amphitheater, the focus of a well is its center, and the studio wanted to create an experience that was outward as well as inward-looking.
By opening up voids between the steps to create a three-dimensional lattice, the public square could be stretched upward, creating over a mile of routes that could be explored in different ways. To create the continuous geometric pattern of the stepwell, with 154 interconnecting flights of stairs, the object had to be self-supporting – a discreet structural solution was required, which did not need additional columns and beams. This was resolved by inserting a steel spine between each pair of staircases, creating a natural division between "up" and "down". The raw welded steel of this structure is exposed to give the object clarity and integrity, and the underside of the staircases is clad in a deep copper-toned metal, setting them apart from the surrounding architecture.
Every element of the Vessel is bespoke, from the joints to the handrails. The 75 huge steel components were produced in Venice by specialist fabricator Cimolai, before being brought from Italy in 6 shipments, carried across the Hudson River by barge, and assembled on site in a process that took 3 years. Yet despite the size of the Vessel, it has been designed at a human scale, to be climbed, explored and enjoyed by New Yorkers and visitors – a simple structure, animated by people and the reflections of the square beneath.

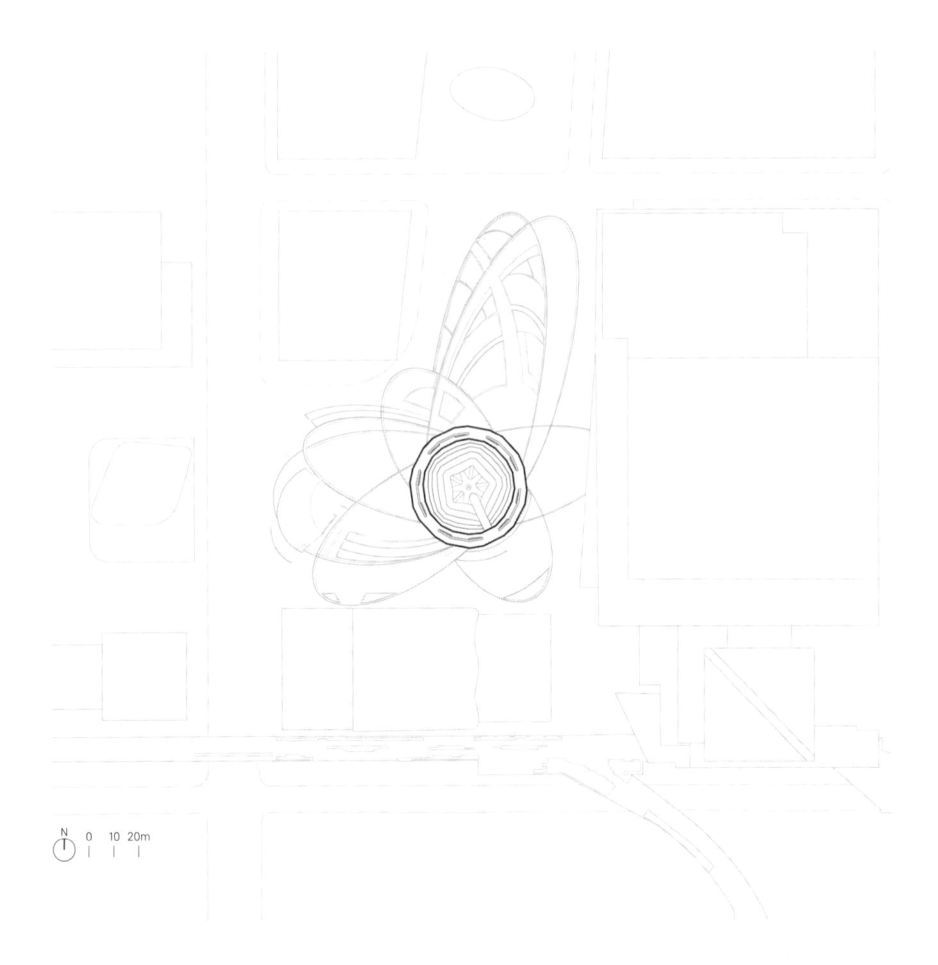
N
0
10
20m

lululemon

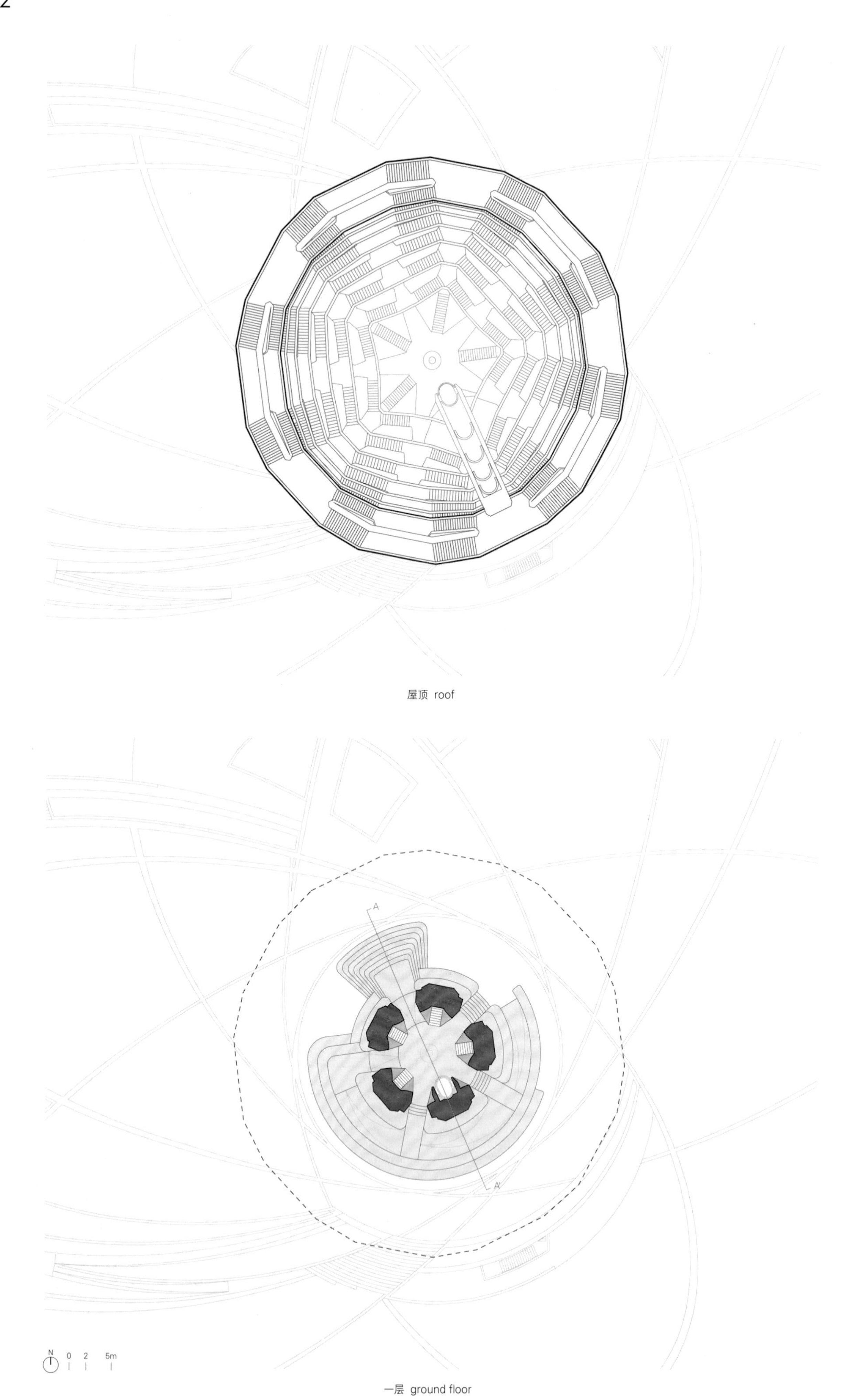

屋顶 roof

一层 ground floor

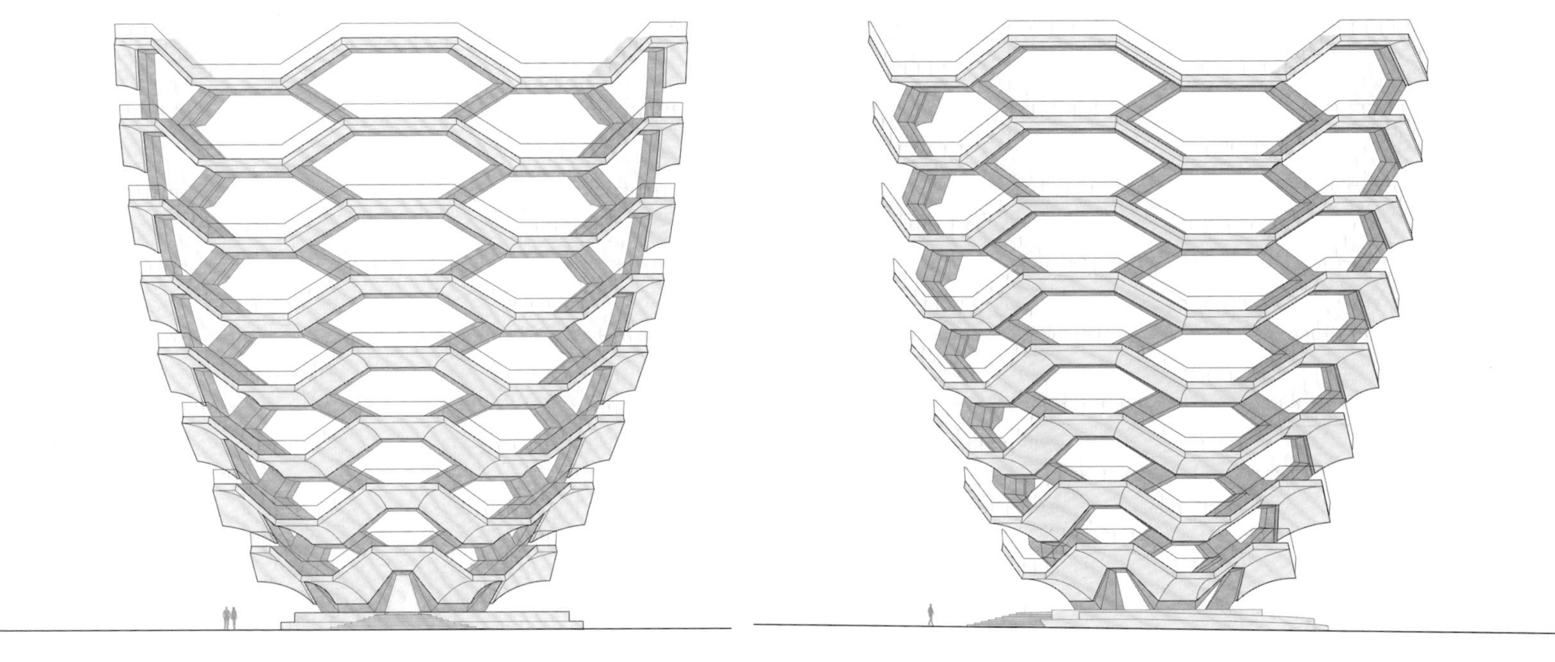

北立面 north elevation

西立面 west elevation

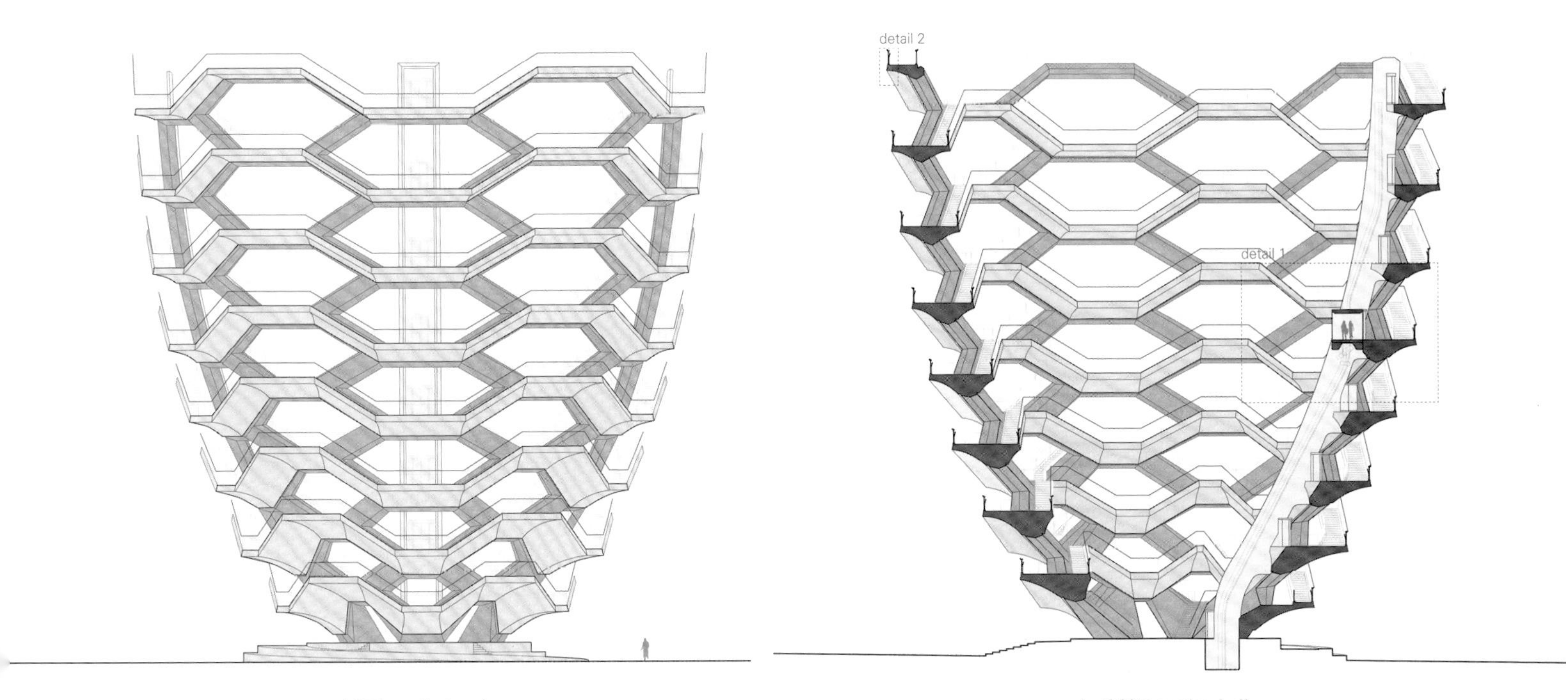

南立面 south elevation

A–A' 剖面 section A-A'

项目名称：Vessel / 地点：Hudson Yards, New York, USA / 建筑师：Heatherwick Studio / 设计总监：Thomas Heatherwick / 团队主管：Stuart Wood / 项目主管：Laurence Dudeney / 项目团队：Charlotte Bovis, Einar Blixhavn, Antoine van Erp, Felipe Escudero, Thomas Farmer, Steven Howson, Jessica In, Nilufer Kocabas, Panagiota Kotsovinou, Barbara Lavickova, Alexander Laing, Elli Liverakou, Pippa Murphy, Luke Plumbley, Ivan Ucros Polley, Daniel Portilla, Jeff Powers, Matthew Pratt, Peter Romvári, Ville Saarikoski, Takashi Tsurumaki / 客户：Related, Oxford Properties Group / 设计工程师：AKTII / 结构工程师：Thornton Tomasetti / 景观设计师：Nelson Byrd Woltz / 注册建筑师：KPF Associates / 钢材承包商：Cimolai / 电梯承包商：Cimolai Technologies / 覆层承包商：Permasteelia / 群体分析：ARUP / 项目管理：Tisham / 建筑面积：2,210m² / 建筑高度：45.7m / 竣工时间：2019.3 / 摄影师：©Michael Moran (courtesy of the architect) - p.15, p.25; ©Getty Images (courtesy of the architect) - p.26~27, p.28~29; ©Francis Dzikowski (courtesy of the architect) - p.20~21; courtesy of Related-Oxford - p.16~17, p.18, p.19

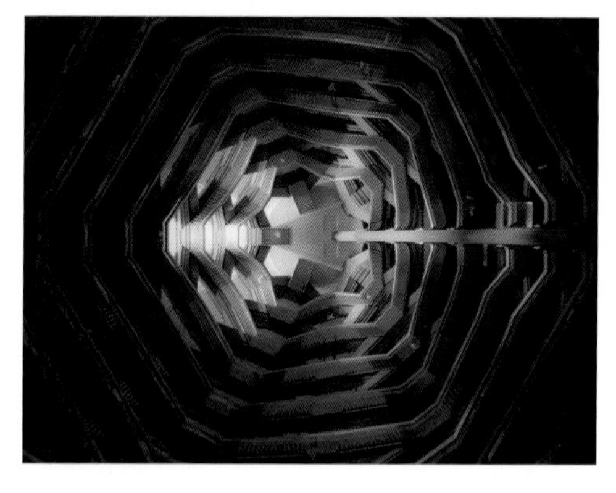

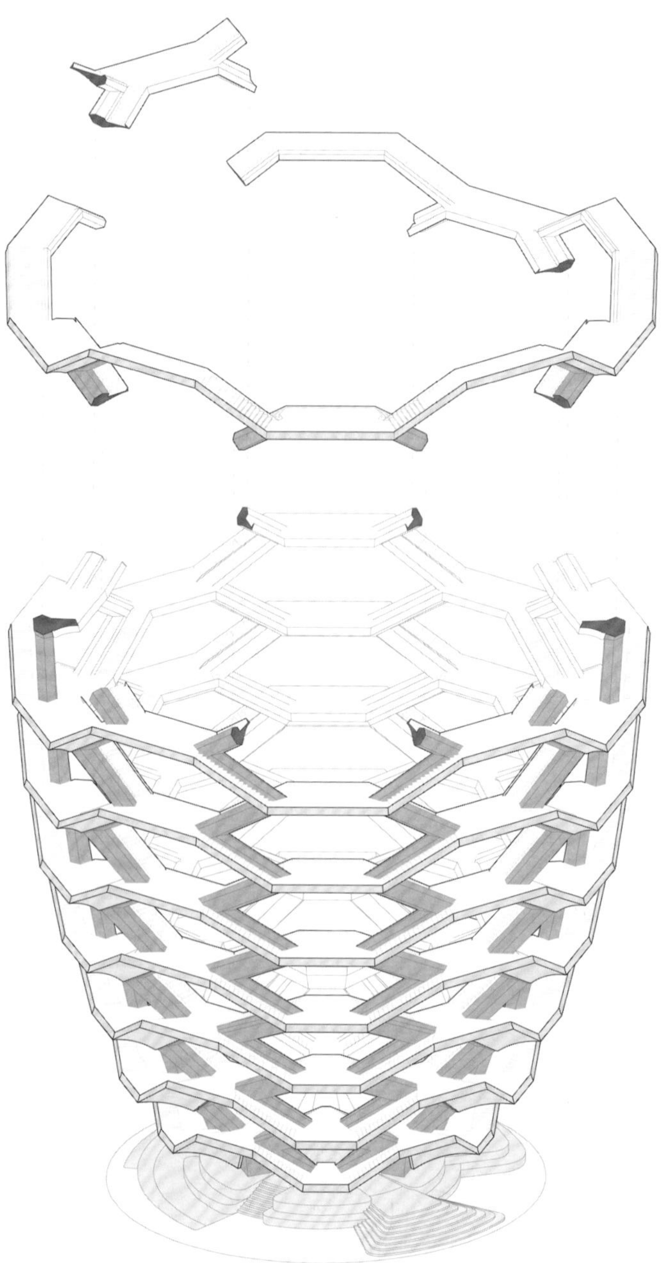

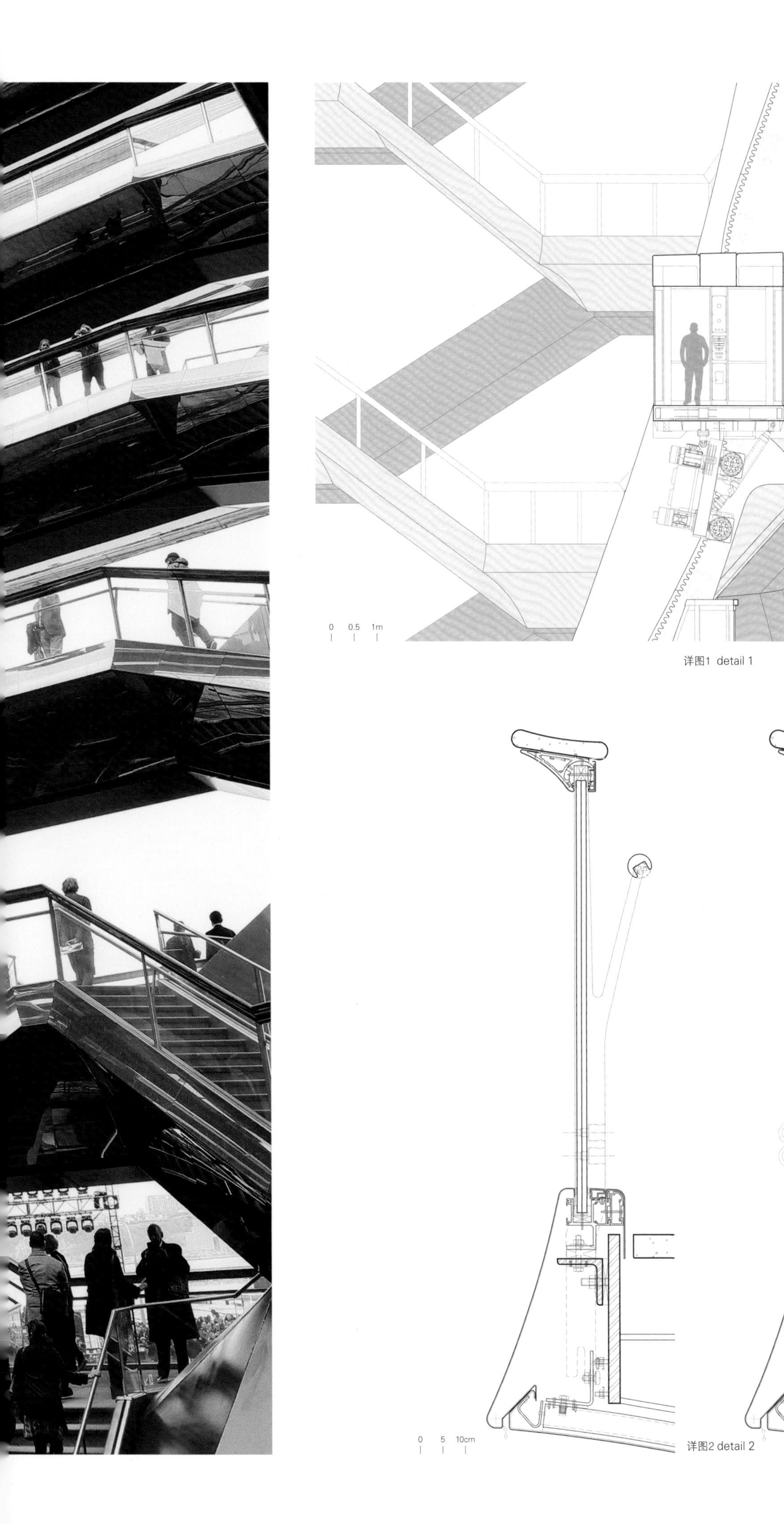

详图1 detail 1

详图2 detail 2

B4
A5
B3
A4
B2
B1
A3
A2

F4
F3
F2
G4
F1
G3
G2

The Shed 大楼

The Shed

Diller Scofidio + Renfro

这个项目致力于为所有观众创造、制作并展示所有不同学科的原创艺术作品。

这座为21世纪艺术家和观众设计的纽约新艺术中心，当它在曼哈顿西侧向公众开放之时，就旨在通过创新的可移动建筑来展示世界上首发的表演艺术、视觉艺术和流行文化作品。

由行业知名的建筑事务所Diller Scofdio + Renfro与罗克韦尔集团合作设计的这座The Shed大楼是一座18500m²的新式建筑，旨在进行实体方面的改造，以支持艺术家最雄心勃勃的设计思想。它的八层基座建筑包括两层画廊空间、多功能剧院、排练空间、创意实验室和有天窗的活动空间；可伸缩的外壳位于基座建筑的上方，可在原地展开，并沿轨道滑动到相邻的广场上，从而使大型表演、装置和活动的建筑面积增加了一倍。

从建筑层面来说，这个项目的灵感来自于游乐宫，这是英国建筑师塞德里克·普莱斯和剧院导演琼·利特伍德在20世纪60年代构思的一个充满影响力但未实现的建筑机器。和它的前身一样，这个项目的开放式基础设施可以为未知的未来提供永久的灵活性，并且对规模、媒体、技术和艺术家不断变化的需求做出反应。

这座建筑物能够通过滚动轨道上的伸缩式外壳来扩大和收缩自身的体积。The Shed项目的动力系统受到了高线公园和西区铁路站的工业历史的启发。通过使用传统的固定结构建筑系统，并采用门式起重机技术赋予外壳活力，该装置能够根据需要适应大型室内项目和露天项目的需求。

展开后，The Shed项目的外壳就形成了一个1600m²的，由灯光、声音和温度控制的大厅，可供多种用途使用。该大厅拥有1200个座位，或2700个站位；基座建筑的两个相邻走廊内可灵活重叠的空间可容纳多达3000名观众。外壳的整个天花板可作为一个剧场平台使用，整个平台都有传动装置和结构功能。朝向广场开放的大门使得项目在开放时能够与北部和东部的公共区域相连接。

当The Shed项目的外壳再次架在基座建筑上时，1810m²的广场将成为开放的公共空间，可用于室外活动的开展；东立面可作为投影的背景，并提供照明和声音支持，同时广场配有用于室外的分布式电源。

暴露在外的斜肋钢框架被包裹在半透明的垫层中，这些垫层由耐用、轻质的基聚合物制成，也被称为聚四氟乙烯（ETFE）。由于中空玻璃的热性能只有其重量的一小部分，因此ETFE就使得光线可以自由通过，并且能够承受飓风的风力冲击。

The Shed is dedicated to commissioning, producing, and presenting original works of art, across all disciplines, for all audiences.

New York City's new art center, designed for 21st-century artists and audiences, will present world premiere works in the performing arts, visual arts, and popular culture in its innovative, movable building when it opens its door to the public on Manhattan's west side.

The Shed's building—an innovative 18,500m²-structure designed by lead architects Diller Scofidio + Renfro, and collaborating architects Rockwell Group—is designed to physically transform, to support artists' most ambitious ideas. Its eight-level base building includes two levels of gallery space, a versatile theater, a rehearsal space, a creative lab, and a skylit event space; a telescoping outer shell can deploy from its position over the base building and glide along rails onto an adjoining plaza to double the building's footprint for large-scale performances, installations, and events.

The Shed takes inspiration, architecturally, from the Fun Palace, the influential but unrealized building-machine conceived by British architect Cedric Price and theater director Joan Littlewood in the 1960s. Like its precursor, the Shed's open infrastructure can be permanently flexible for an unknowable future and responsive to variability in scale, media, technology, and the evolving needs of artists.

The building is able to expand and contract by rolling the telescoping shell on rails. The Shed's kinetic system is inspired by the industrial past of the High Line and the West Side Railyard. Through the use of conventional building systems for the fixed structure, and adapting gantry crane technology to activate the outer shell, the institution is able to accommodate large-scale indoor and open-air programming on demand.

PARK

THE SHED
OPENS SPRING

When deployed, the Shed's shell creates a 1,600m² light-, sound-, and temperature-controlled hall that can serve a variety of uses. The hall can accommodate an audience of 1200 seated or 2700 standing; flexible overlap space in the two adjoining galleries of the base building allows for an expanded audience of up to 3000 in the hall. The shell's entire ceiling operates as an occupiable theatrical deck with rigging and structural capacity throughout. Large operable doors on the plaza level allow for engagement with the public areas to the north and east when open.

When the Shed's shell is nested over the base building again, the 1,810m² plaza will become open public space that can be used for outdoor programming; the eastern facade can serve as a backdrop for projection with lighting and sound support. The plaza is equipped with distributed power supply for outdoor functions.

The exposed steel diagrid frame is clad in translucent pillows of durable and lightweight teflon-based polymer, known as ethylene tetrafluoroethylene (ETFE). With the thermal properties of insulating glass at a fraction of the weight, ETFE allows light to pass through and can withstand hurricane-force winds.

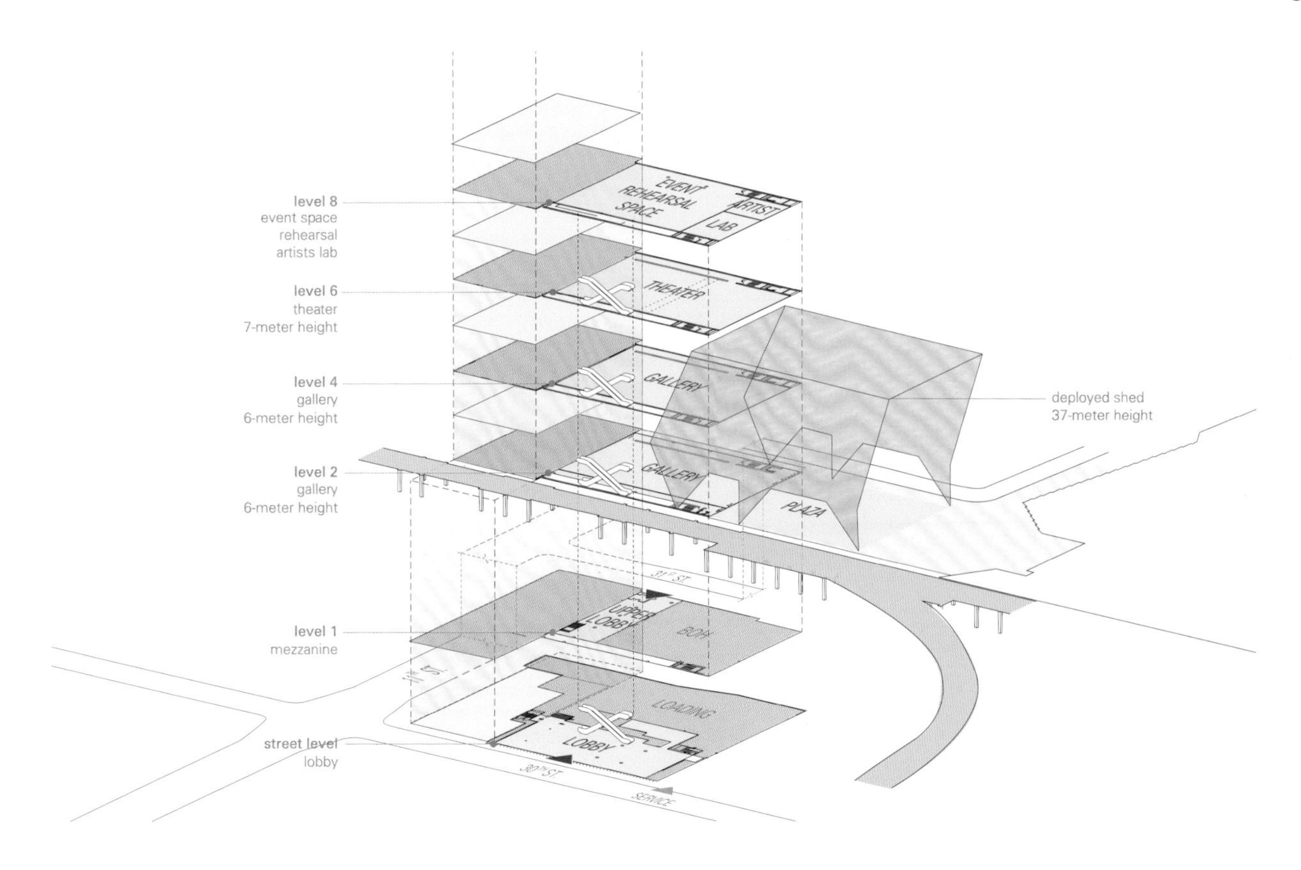
level 8
event space
rehearsal
artists lab
level 6
theater
7-meter height
level 4
gallery
6-meter height
level 2
gallery
6-meter height
level 1
mezzanine
street level
lobby
deployed shed
37-meter height
EVENT
REHEARSAL
SPACE
ARTISTS
LAB
THEATER
GALLERY
GALLERY
PLAZA
UPPER
LOBBY
BOH
LOADING
LOBBY
SERVICE

BAKER
APR 6
APR 6 MAY 30
APR 6
RICHTER
PART

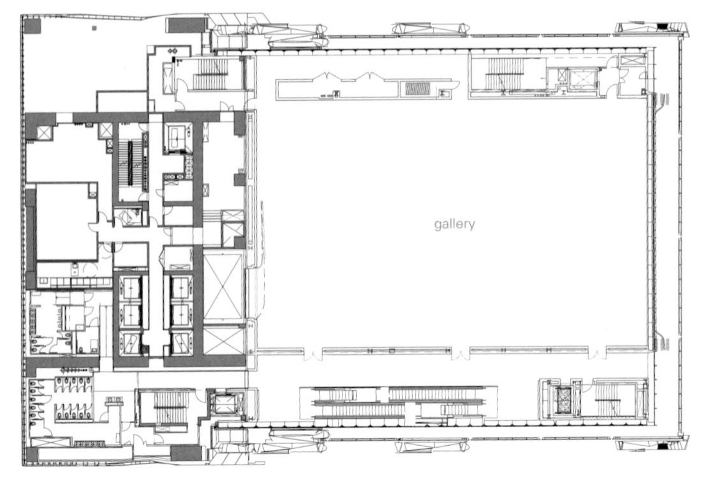

四层 level 4 floor

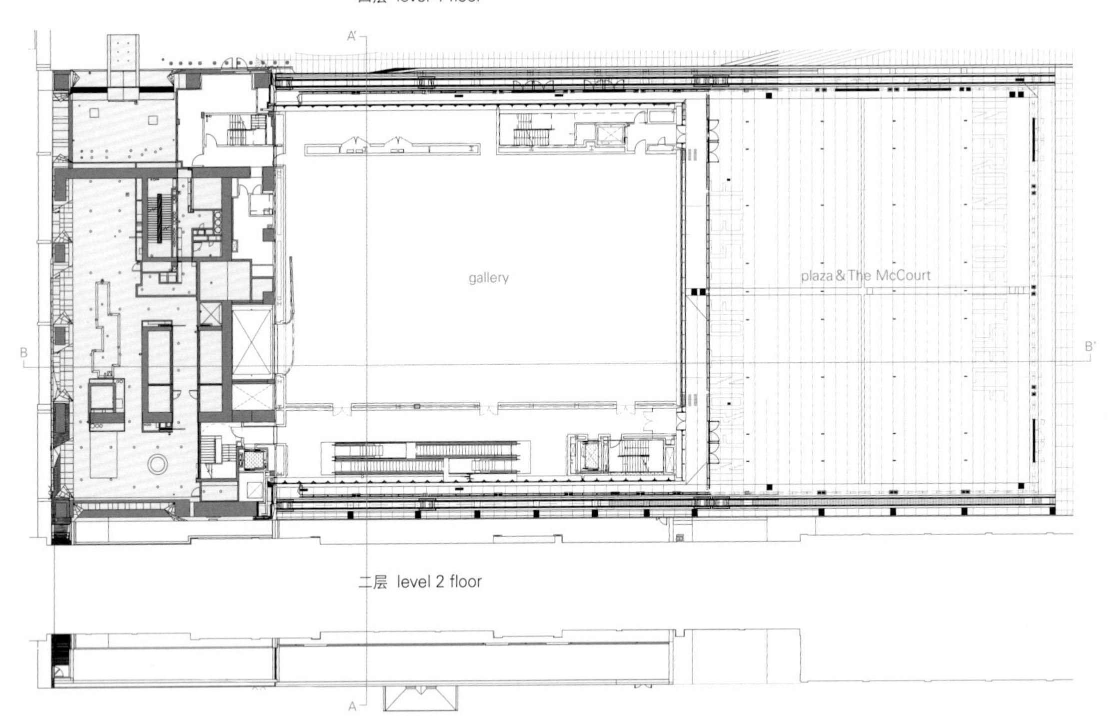

二层 level 2 floor

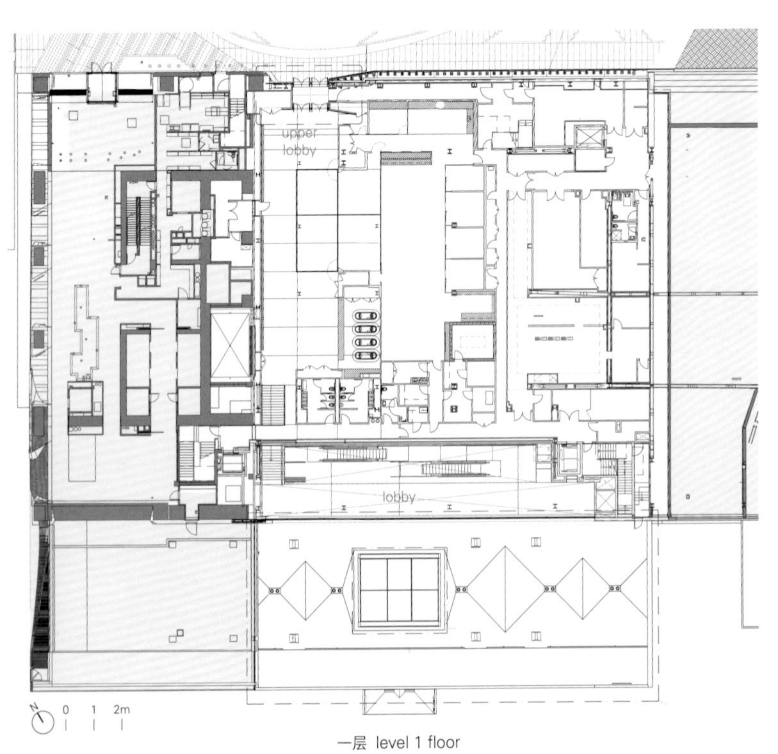

一层 level 1 floor

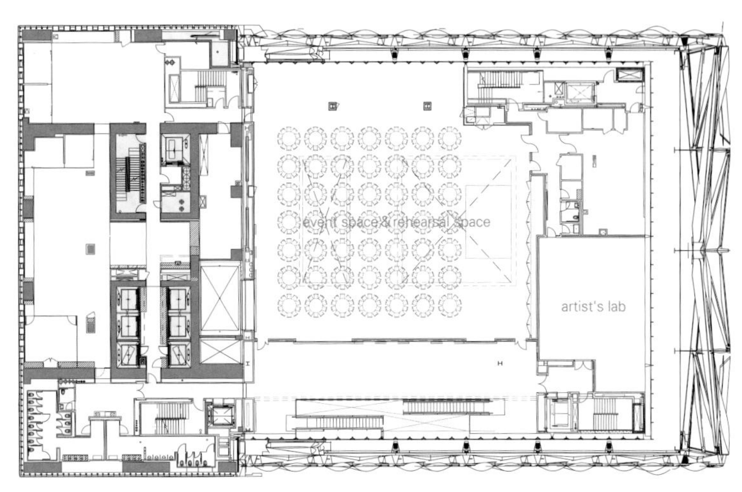

八层 level 8 floor

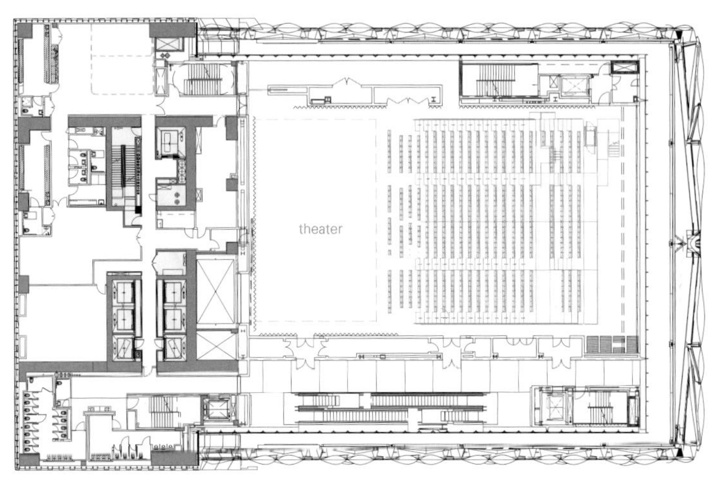

六A层 level 6A floor

项目名称：The Shed
地点：West 30th Street between
10th and 11th Avenues, New York, United States
主持建筑师：Diller Scofidio + Renfro
合作建筑师：Rockwell Group
结构工程师：Thornton Tomasetti
动力系统顾问：Hardesty & Hanover
机电&消防顾问：Jaros Baum & Bolles
能源模型顾问：Vidaris
照明顾问：Tillotson Design Associates
视听声效顾问：Akustiks
剧场设计顾问：Fisher Dachs
钢结构制造商：Cimolai
施工经理：Sciame Construction, LLC
客户：The Shed
总建筑面积：200,000m²
设计时间：2008
施工开始时间：2015
开始时间：2019.4.5
摄影师：
©Brett Beyer (courtesy of The Shed) - p.32~33, p.41
©Iwan Baan (courtesy of The Shed) - p.31, p.34, p.36~37
©Timothy Schenck (courtesy of The Shed) - p.40, p.42~43

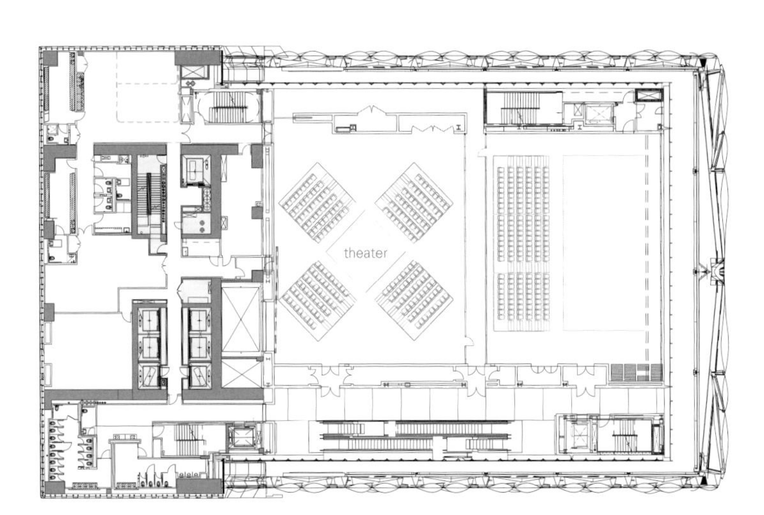

六B层 level 6B floor

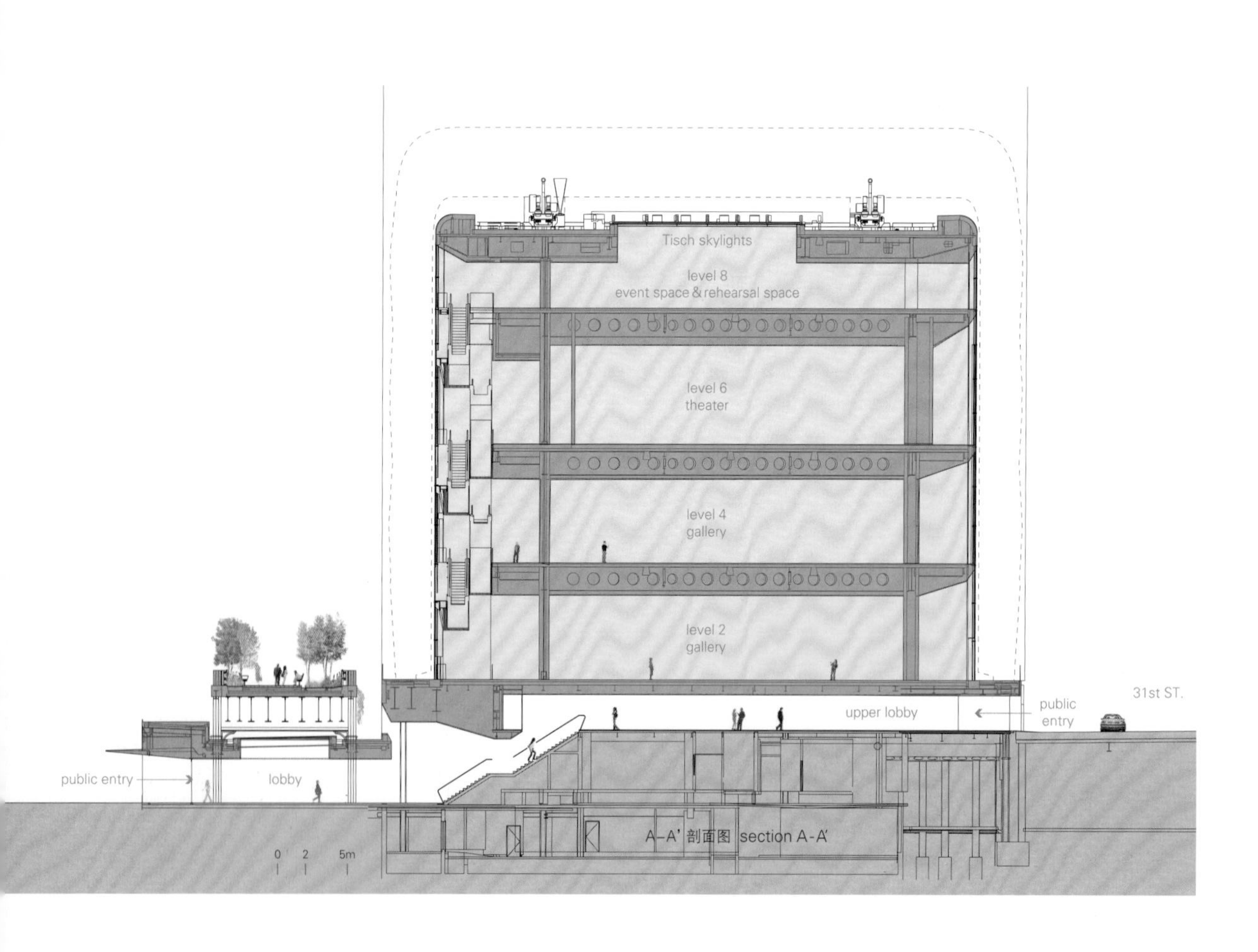

A–A' 剖面图 section A-A'

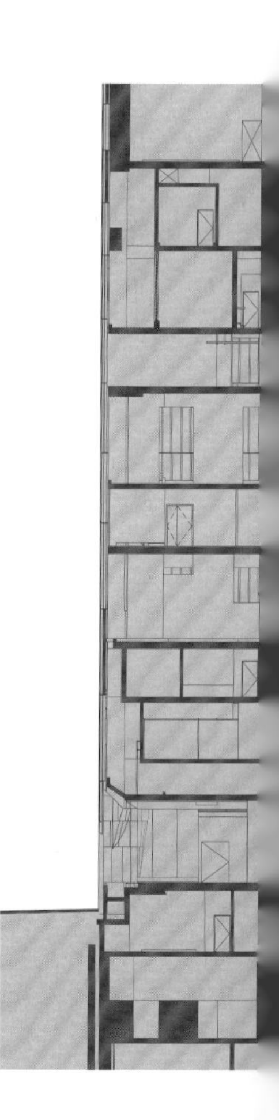

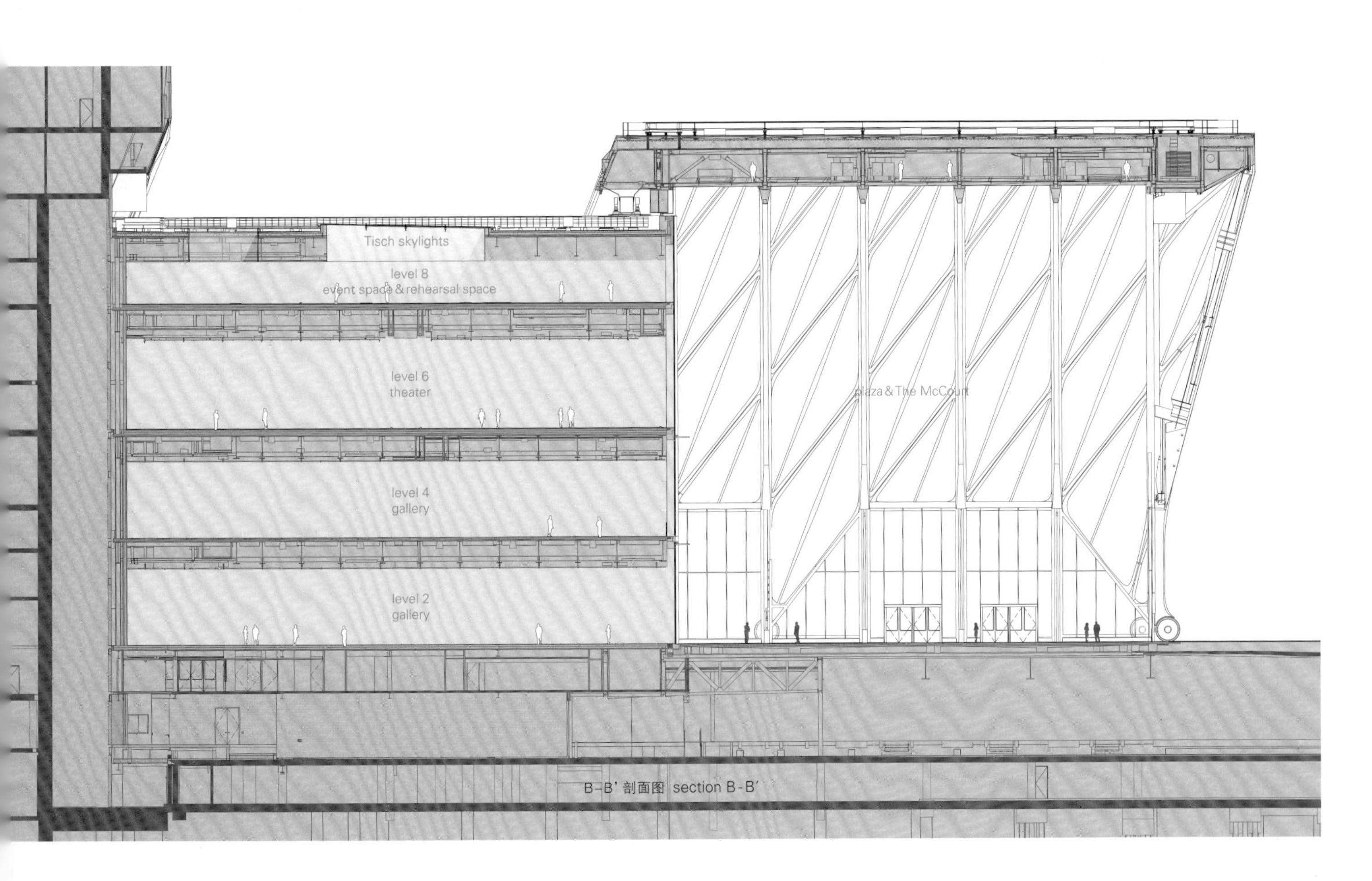

B-B' 剖面图 section B-B'

公共图书馆的社会资本建设

Building The In The Publ

本文着重论述了当代公共图书馆的演变过程，阐述了近几十年来影响它们作为"图书宝库"角色的变化。与20世纪上半叶现代图书馆建筑的僵化相比，现代设计逐渐向城市公共空间开放，成为"第三空间"概念的中心。"第三空间"是指除家和工作场所以外的地方，具有包容性和社交性。在此期间，图书馆的关键作用在社会资本的增强过程中变得越来越重要，设计师的努力重心从打造专门收集和检索印刷书籍和手稿的建筑物的功能影响，转移到了更广泛、专业性不太强的系列功能方面，主要围绕知识共享和文化推广与

This essay focuses on the process of evolution of contemporary public libraries, describing the changes that affected their role as "repositories of books" in the last decades.
By comparison with the architectural rigidity of modern library, as it resulted during the first half of the 20th Century, contemporary designs progressively opened towards urban public spaces, becoming central in the notion of "third place": places other than the home or the workplace, inclusive and sociable. During this period, the pivotal role of libraries in the enhancement of social capital became increasingly more important, shifting the efforts of designers from the functional implications of buildings devoted to the collection and retrieval of printed books and manuscripts, to a broader and less specialized series of functions, mainly

Oodi中央图书馆_Oodi Central Library/ALA Architects
卡尔加里新中央图书馆_Calgary's New Central Library/Snøhetta + DIALOG
公共图书馆的社会资本建设_Building the Social Capital in the Public Libraries/Davide Pisu

Social Capital
ic Libraries

生产的理念而展开。

在这一过程中，图书馆已成为一种公共设施，拥有培育社区关系和促进人与人之间互动的潜力，书籍只是这其中信息传递和交流的各种手段之一。通过对一系列当代图书馆项目的分析，本文阐述了伴随而来的趋势和变化，并以此解释了上述存在的现象。文章认为，设计师对这些变化非常敏感，当代项目对这些趋势表现出极大的认知度，文章也着重强调了设计师在此设计过程中所采取的策略。

orbiting around the idea of shared knowledge and cultural promotion and production.
In this process, the library has become a public amenity, potentially nurturing communities and fostering human interaction, in which books are only one of the various means through which information is transmitted and communicated. These trends and changes are explained through the analysis of a series of contemporary library projects by which the aforementioned phenomena are explained. The article argues that designers are extremely sensitive to such changes, and contemporary projects show a great awareness of these trends, emphasizing the strategies adopted by the designers.

公共图书馆的社会资本建设

Building the Social Capital in the Public Libraries

Davide Pisu

当人们不在家或工作场所时，他们的时间都花在哪里了呢？咖啡馆、书店，健身房，成功的公共空间是一个自由出入、自由参与、无排斥标准和亲疏观念的领域，上述这些特质吸引人们前来，并鼓励他们在此进行互动。根据美国社会学家雷·奥登伯格的著作《绝好的地方》[1]，这些"第三空间"（第一个是家庭空间，第二个是工作场所）在建设社区的活力方面起到了重要的作用，对营造一个安全、具有活力的公共领域产生了至关重要的作用。

公共图书馆是这个领域的重要组成部分。尽管它们显而易见的特性是作为存储、知识和检索信息的仓库，但在近三十年来，它们的作用一直在放大。由于与万维网有关的新媒体方式的不断出现，公共图书馆作为知识和信息接入点的作用已经减弱，取而代之的是个人设备（例如，计算机、智能手机和平板电脑）的分布式网络。然而，公共图书馆作为社会交往的场所，逐渐获得了新的生命力，面向外界开展了一系列形式多样的活动：纸质图书的获取、编目和保存只是其中之一。正如Putnam和Feldstein[2]所说，公共图书馆在提高社区社会资本的过程中发挥着关键作用。

尽管我们可以假设这些变化主要与图书馆工作方式的非物质方面有关，但从20世纪90年代开始，图书馆的设计就已经开始发生了变化，作为对这种现象的回应。尤其是当图书馆开始以前所未有的方式与公共空间进行接触。这种参与包括从外部到内部都使用了连续的地板表面和透明的立面，来强化对街道、广场和公园的开放程度。但有时，这种回应以一种更微妙、更有

When not at home or in the workplace, where do people spend their time? Cafés, bookstores, gymnasiums and successful public spaces define a domain where freedom of access and participation, the absence of criteria of exclusion and the idea of intimacy all combine to attract people and encourage their interaction. According to American sociologist Ray Oldenburg's famous book, *The Great Good Place*[1], these "third places" – whereby the first is the domestic space and the second the place for work – have a fundamental role in building the vitality of a community, and are of absolute importance for the very existence of a safe, common and lively public domain.
Public libraries are an important part of this domain. Despite their apparent nature as repositories for storage, knowledge and retrieval of information, their role has been expanding during the last three decades. Due to the growing emergence of new media related to the world wide web, the role of public libraries as an access point to knowledge and information has diminished in favor of the distributed network of personal devices such as computers, smartphones and tablets. However, public libraries gradually acquire a new life as places of social interaction, open to a series of diverse activities of which the acquisition, cataloguing and conservation of printed books is just one. As Putnam and Feldstein[2] argue, public libraries have a key role in the process of enhancing the social capital of a community.
While one can assume that these changes pertain mostly to the immaterial aspects of libraries' modus operandi, a shift in the design of libraries, starting in the 1990s, took place as a response to this phenomenon. In particular, the library started engaging with the public space in an unprecedented manner. This engagement may have consisted of an augmented openness towards streets, squares and parks, with continuous

效的方式展开。这其中包括把图书馆的内部空间设想为一个公共空间，在一个充满"广场、公园、纪念楼梯、咖啡馆、精品店"的空间中，"访客变成了漫游者"。卡尔加里的新中央图书馆（第76页）由Snøhetta和DIALOG设计，位于一个极其复杂的城市区域内，与轻轨交通线的柔和曲线交叉。根据这种情况，设计师们选择利用这种几何图形的设计方法，在菱形体块中运用镜像的方法体现它的存在。入口是通过在地面上建造一个拱形的具有纪念意义的天篷建造而成的，而整个立面由玻璃和白色铝板组成，天篷覆盖着红杉木板，营造出室内空间的迎宾气氛，并强调了主入口的存在感。整座建筑都建在一个裙楼上，在裙楼上有一系列的楼梯，让公众可以坐下来体验周围的公共空间。一旦进入其中，参观者将面对一个全高的空间。顶部安装了一个巨大的采光天窗，可以即时显示出图书馆的尺寸和几何特征。环绕中庭的螺旋楼梯采用了简单的几何设计原理，使得使用者可以毫不费力地在建筑内部确定方位。但尽管如此，内部空间的特色是充满了一系列的错位和空间效果，这样一来，图书馆内部的不同区域都会给你带来独特的视觉体验，人们也能在各个地方都欣赏到不一样的城市景观。

因此，图书馆是一个统一的空间，其交通流线和观景位置是主要的形式标准。尽管拥有这种统一性，设计师们还是设法为各种活动创造了一系列不同寻常的独特安排。例如，在一层设置了一间礼堂，在夹层设置了一个咖啡厅，另外还沿着木楼梯一路上升布置了儿童图书馆、技术实验室和作家区域等一系列房间和区域。尽管规模如此庞大，卡尔加里新中央图书馆的内部空

floor surfaces from the outside to the inside and transparent facades. But sometimes, in a more subtle and effective fashion, it consisted of an interior space of the library conceived as a public space itself, where "The visitor becomes a flaneur" in a space filled with "plazas, parks, monumental stairways, cafés, boutiques"[3]. Calgary's New Central Library (p.76), designed by Snøhetta and DIALOG is located within an extremely complex urban area, crossed by the soft curve of the light rail transit line. Drawing upon this condition, the designers chose to exploit this geometry, mirroring it in a lozenge-shaped block. The entrance is obtained by carving a monumental canopy in the form of an arch at the ground floor, and while the whole facade is composed of glass and white aluminum panels, the canopy is clad in red cedar planks that anticipate the welcoming atmosphere of the interior space and emphasize the main access. The entire building rests on a podium, on which a series of stairs encourage the public to sit and to experience the surrounding public space. Once inside, the visitors are confronted with a full-height space, terminating in a vast skylight that displays the dimensions and geometric features of the library instantaneously. The simple geometric principle of the spiral staircase running all around the atrium, allows the users to effortlessly orientate inside the building, but in spite of that, the interior space is characterized by a series of displacements and spatial effects, allowing peculiar points of view from distinct areas within the library and towards the cityscape.
The library is thus a unified space, where circulation and points of view are the main formal criteria. Despite this unity, the designers manage to create a series of unique and peculiar arrangements for various kinds of activities. For example, an auditorium is located at the ground floor, a cafe at the mezzanine, and a series of rooms and areas ranging from the children's library to technology labs and writers' spots are grafted along

卡尔加里的新中央图书馆，加拿大
Calgary's New Central Library, Canada

间仍然传达了一种亲密和熟悉的感觉。这反过来又鼓励用户更多地在室内享受美好时光，并亲自投身到可以在图书馆举办的大量公共活动中，这些活动之所以可以在这里举办，也是因为这是一个丰富而又高品质的场所。

2012年举办的一场建筑设计竞赛吸引了544名设计师参与，最终产生了赫尔辛基Oodi中央图书馆（第50页）这个项目。ALA建筑师事务所的设计最终获奖。这是一家位于赫尔辛基的芬兰建筑事务所，专门从事公共建筑和文化建筑设计。图书馆位于芬兰议会大厦前的Kansalaistori广场，该广场位于城市中心。设计展示了对图书馆作为城市纪念碑这一作用的深刻诠释，将图书馆的室内环境投射到了外部的广场和城市的公共空间之中。

与卡尔加里中央图书馆完全不同的是，Oodi中央图书馆的特色是一个盒子状的几何体，跨度超过100m，有一系列的错位和变形，这似乎是对原本简单的形式进行自由操作的结果。这些变形中的第一种限制了底层的体块，并将其上部的楼层延伸成一个天篷，欢迎公众大批涌入。同样，位于卡尔加里的Snøhetta设计的图书馆天篷是用木材覆盖的，从而烘托了室内空间的温暖气氛。地面层被视为广场的延伸，并具有一系列公共功能，包括服务设施、画廊和展览空间、多功能厅、咖啡厅和电影院。这个空间被一个令人印象深刻的木制拱顶所覆盖，将整座建筑的长度都包含在内，它传递了一种纪念的意味，能迅速地吸引游客进入建筑的中心来体验。

the ascending path of the wooden staircase. Despite the monumental scale, the interior space of Calgary's new Central Library communicates a sense of intimacy and familiarity. This in turn encourages users to spend time inside, and involve themselves in the numerous communal activities that, due to the richness and quality of the place, can be performed in the library facilities.

Oodi Central Library (p.50) in Helsinki, was the result of a competition held in 2012 that attracted 544 designers. The winning design was that of ALA Architects, a Finnish firm based in Helsinki, specializing in public and cultural buildings. The library is located in the Kansalaistori square in front of the Finnish parliament, a place central to the city. It demonstrated a deep understanding of the role of the library as urban civic monument, projecting its interior towards the square and the city's public space.

Quite different from Calgary Central Library, Oodi is characterized by a box-like geometry, spanning over 100m, with a series of displacements and distortions that seem as the result of the free manipulation of an originally simple form. The first of these deformations constrains the block at the ground floor and extends its upper floor into forming a canopy, inviting the public to flow in. Similarly, the canopy of Snøhetta's design at Calgary is clad in timber, and anticipates the warm atmospheres of the interior spaces. The ground floor is conceived as an extension of the square, and accommodates a series of public functions, including services, a gallery and exhibition space, a multi-purpose hall, a café and a cinema. The space is covered by an impressive wooden vault, comprising all the length of the building, that transmits a sense of monumentality and immediately welcomes the visitors into the heart of the building.

The first floor is the most introverted space of the library, with a lower ceiling than the others. It hosts a

Oodi中央图书馆，芬兰
Oodi Central Library, Finland

一层是图书馆最内向型的空间，天花板比其他楼层都低。它拥有各种各样的办公室、实验室、制造区域和工作区。楼层所有设施均邀请用户参与文化方面和材料方面的制作。第二层也是最后一层，则完全留给了书籍，并被构想为一个城市公共空间：除了书架之外，建筑的整个长度上散布着树木、块状的服务设施、长椅、台阶和露台。在一些小方格中，人们或嬉戏，或就坐又或者看看书。立面的整个高度都是透明的，可以与周围的环境保持稳定的视觉关联。

这两个最近完成的图书馆项目重新审视了自信息技术革命开始以来，公共图书馆所拥有的更广泛但仍万变不离其宗的作用。通过这两个项目，我们还可以洞察这些公共建筑的未来，就是要重新满足市民对其存在的需求。它们是文化和知识的丰碑，促进社会资本的产生，并培育了更富有、更充满活力的社区。

在这种情况下，建筑师和设计师的工作展现了一种趋势，即图书馆成为社区的社会和政治参与的空间——在那里，信息获取的空间与用于制作媒体的实验室以及鼓励积极参与和辩论的演讲厅相结合——这是今天这个全民参与的时代的宣言，向城市展示这里是知识共享的灯塔。

variety of offices, laboratories, a makerspace and working areas. All the facilities of the floor invite users to engage with cultural and material productions. The second and also the last floor is completely dedicated to books, and is conceived as an urban public space: scattered along the full length of the building, aside from the bookshelves, are trees, block-like services, benches, steps and a terrace. There are small squares in which it possible to play, or simply sit and read. The full height of the facade is transparent, allowing a constant visual relationship with the surroundings.

These recent two library projects reflect the wider yet still central role that public libraries have invested in becoming since the beginning of the information technology revolution. They also allow for an insight into the future of such public buildings, renewing the civic need for their presence. They stand as a monument to culture and knowledge, fostering social capital and allowing for a richer, more vibrant and vital community.

Architects' and designers' work, in this case, reveals a trend in which the library becomes a space for the social and political engagement of communities —where the spaces for access to information are paired with labs for the production of media and lecture halls that encourage active participation and debate— and a sort of manifesto of the age of participation, that displays itself to the city as a beacon of shared knowledge.

1. Ray Oldenburg, *The Great Good Place: cafes, coffee shops, bookstores, bars, hair-salons and other hangouts at the heart of a community* (New York: Marlowe and Company, 1989)
2. Robert D. Putnam and Lewis M. Feldstein, *Better Together: restoring the American community* (New York: Simon and Schuster, 2003)
3. The original quote was referred to OMA's unbuilt jussieu campus library, from Rem Koolhaas and Bruce Mau, *S,M,L,XL* (New York: Monacelli press, 1998), 1320.

Oodi 中央图书馆

Oodi Central Library

ALA Architects

赫尔辛基的Oodi中央图书馆所在的位置具有非凡的意义：它既面向着芬兰议会大厦的台阶，又面向Kansalaistori广场上那些主要的市政机构。该位置象征着政府和民众之间的关系，提醒人们《芬兰图书馆法》对图书馆的要求，即促进终身学习，并提倡公民意识、民主和言论自由。有趣的是，书籍只占图书馆空间的三分之一。通过减少现场的书籍存储，并向图书馆的用户咨询他们是如何接触文化事物的，设计师能够引入更多的文化设施。这代表了芬兰图书馆在借阅书籍之外对于开展更广泛的服务方面所进行的尝试。

Oodi图书馆的空间概念包括三个公共楼层；它是一座可以住人的不对称钢桥结构，横跨在直径大于100m的露天地面之上。整个结构由两个巨大的钢拱支撑，钢拱再通过钢筋混凝土板张拉在一起。这一创新的解决方案提供了灵活的无柱室内空间，并考虑到了将来可以在此建造地下隧道。通过二元钢桁架支持悬臂式阳台和屋顶天篷，设计师创造出一个大型的公共平台，使室外公共空间翻倍，并正对着广场，一览无余。Oodi图书馆的一层将Kansalaistori广场延伸到其内部，提供了一个非商业用途的室内公共空间，每天向所有人开放。Kino Regina是芬兰国家视听研究所的电影院的名字，它位于一层，这里还有一家西餐厅，餐厅的座位在夏季可以扩展到广场上。中间的楼层，或称"阁楼"，由可灵活布置的房间组成，这些房间排列在桥梁桁架之间的隐秘角落里。多功能房间的设计目标是既能举办喧闹的活动，也能举办安静的活动，制作人空间和视听录音室也一并提供。"书的天堂"位于顶层，这是一个巨大的开放式景观，空间上方有一个波浪起伏的、如云层一般的白色天花板，上面有若干个圆形的采光天窗。这种效果在现代主义图书馆和21世纪科技之间创造了协同效应。这里宁静的氛围吸引游客阅读、学习和思考。从这里，游客可以畅通无阻地360度欣赏城市中心的全景，或登上露台俯瞰Kansalaistori广场。

维护和后勤空间位于地下室。这些设施在公众可以到达的楼层中都尽量缩减至最少，以便为用户创建一种开放的、易于接近的体验。除了核心业务之外，设计还提供了广泛的辅助设施，包括提供赫尔辛基市乃至整个欧洲的公共信息。其他设施包括社区厨房、展览空间和位于图书馆内外的儿童游乐场。

根据当地的气候条件，设计师采用了当地材料来建造Oodi图书馆。建筑立面覆盖着33mm厚的芬兰云杉木板，形成了一条将建筑物向外延伸的宽阔曲线，在广场上方创造了一个天篷，将室内外空间融合在一起，为公共活动创造了一个有遮挡的区域。通过采用精确算法辅助参数化3D设计方法，设计师为整座建筑创造出了复杂的曲面几何形

体。建筑外围护结构采用了被动能耗原理，因此该项目几乎可以被称为零能耗建筑（nZEB）。对立面性能的详细分析为环境解决方案提供了依据，并允许项目团队得以采用最小化的机械控制系统。丰富的日光通过玻璃幕墙照亮整个公共区域，提高图书馆作为多功能市民广场的内部空间的品质。

Oodi Central Library, Helsinki, sits in a significant location: it faces the steps of the Finnish parliament – the Eduskuntatalo – as well as the major civic institutions which flank Kansalaistori Square. The site symbolizes the relationship between the government and the populace, acting as a reminder of the Finnish Library Act's mandate for libraries to promote lifelong learning, active citizenship, democracy, and freedom of expression. Interestingly, books only fill one third of this library's space. By reducing on-site storage, and consulting library users on how they access culture, architects were able to introduce additional cultural facilities. This is representative of broader experimentation among Finnish libraries to offer new services beyond loaning books.

Oodi's spatial concept comprises three public floors; it is built as an inhabited, asymmetrical steel bridge spanning over 100m over the open ground floor space. The stucture is supported by two massive steel arches tensioned together with a reinforced concrete slab. This innovative solution provides flexible column-free interior spaces and allows for the future construction of a subterranean road tunnel. Secondary steel trusses, supporting the cantilevered balcony and roof canopy, create a large public terrace which doubles the public outdoor space and looks directly over the square.

Oodi's ground floor extends the Kansalaistori square into its interior, providing a non-commercial interior public space open daily to all. Kino Regina, the National Audiovisual

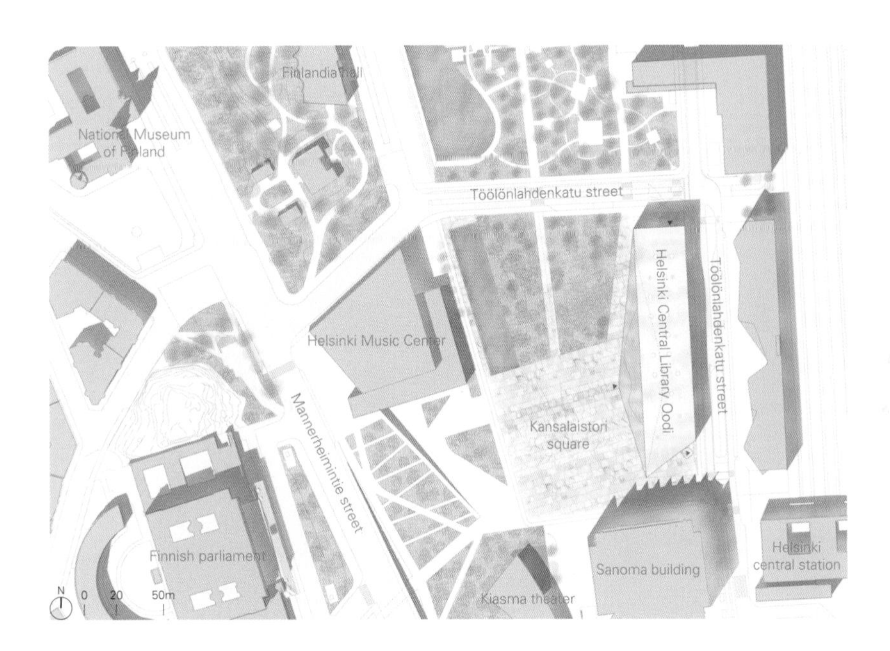

西立面 west elevation

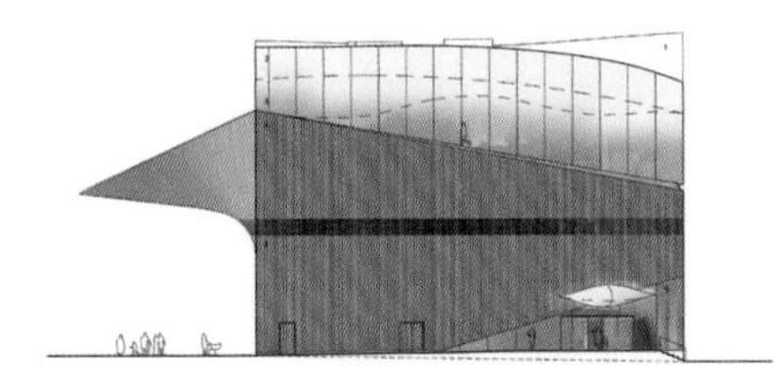
南立面 south elevation

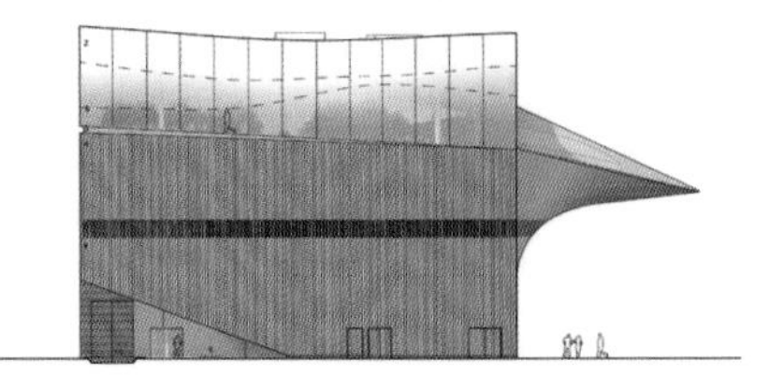
北立面 north elevation

项目名称：Oodi Central Library / 地点：Töölönlahdenkatu 4, 00100 Helsinki, Finland / 建筑师：ALA Architects / ALA竞赛团队：Juho Grönholm, Antti Nousjoki, Janne Teräsvirta and Samuli Woolston with Aleksi Niemeläinen, Jussi Vuori and Erica Österlund, as well as Willem Barendregt, Martin Genet, Vladimir Ilic, Tiina Liisa Juuti, Julius Kekoni, Auvo Lindroos, Pekka Sivula, Pekka Tainio and Jyri Tartia / ALA执行团队：Juho Grönholm, Antti Nousjoki, Janne Teräsvirta (until 2015) and Samuli Woolston with Niklas Mahlberg, Jussi Vuori, Tuulikki Tanska, Tom Stevens, Nea Tuominen, Pauliina Rossi, Anna Juhola and Miguel Silva, as well as Michal Bala, Marina Diaz Garcia, Jyri Eskola, Zuzana Hejtmankova, Harri Humppi, Mette Kahlos, Anniina Kortemaa, Felix Laitinen, Malgorzata Mutkowska, T. K. Justin Ng, Marlène Oberli-Räihä, Olli Parviainen, Alicia Peña Gomez, Anton Pramstrahler, Jack Prendergast, Akanksha Rathi, Niina Rinkinen, Mikael Rupponen and Pekka Sivula / 竞赛阶段合作伙伴：Arup – energy technical specialist, mechanical engineering, structural engineering and facade engineering; VIZarch – renderings; Klaus Stolt – scale models / 执行阶段合作伙伴：YIT – main contractor; E.M. Pekkinen – contractor for excavation work and basement construction; Ramboll CM – project management; Ramboll Finland – structural engineering, HVAC, energy technology; Granlund – theater technology; Rejlers – electrical engineering and AV consulting; Finnmap Infra & Sipti Infra – geo planning; Helimaki Acoustics – acoustical engineering; Gravicon – BIM coordinator; Palotekninen insinööritoimisto Markku Kauriala – fire safety; Pöyry Finland – traffic planning; Saircon – kitchen design; VIZarch – renderings; StoltModels – scale models / 客户：City of Helsinki / 功能：library facilities, meeting rooms, group working space, maker space, living lab, recording studios, photography studio, editing rooms, office space, café, restaurant, movie theater, auditorium, multi-purpose hall, exhibition facilities, information booths / 总建筑面积：17,250m² / 造价：98,000,000 € / 施工开始时间：2015 / 竣工时间：2018.12 / 摄影师：©Tuomas Uusheimo (courtesy of the architect)

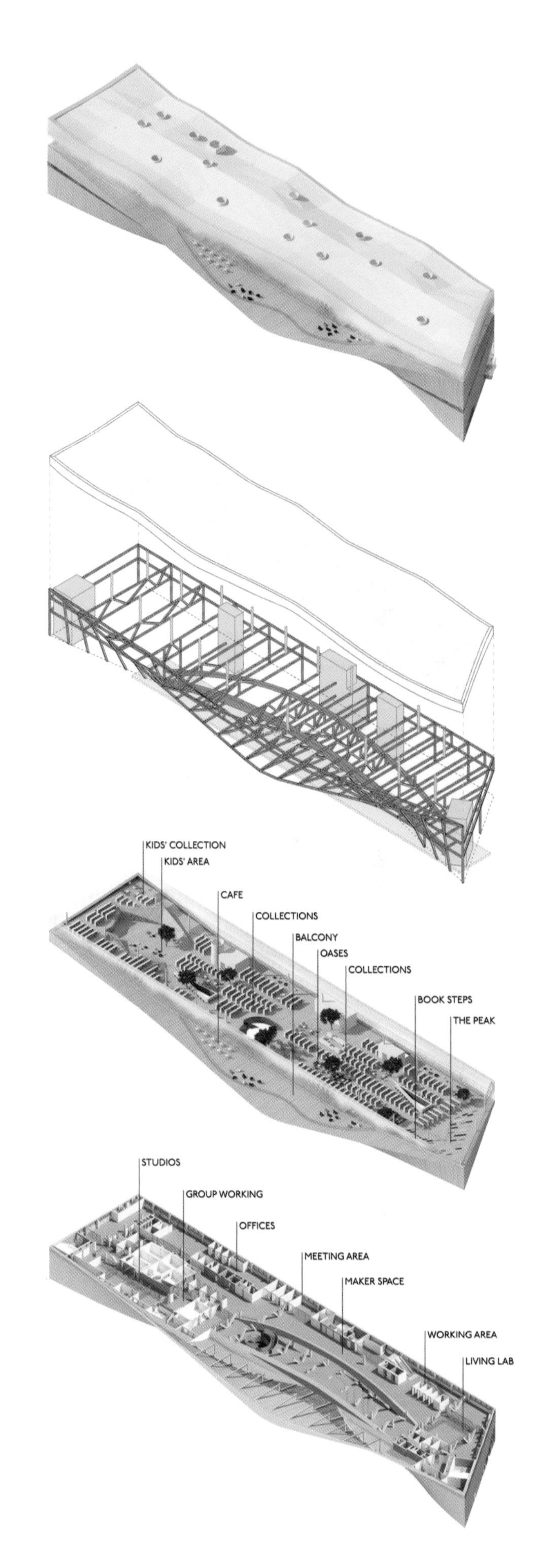
KIDS' COLLECTION
KIDS' AREA
CAFE
COLLECTIONS
BALCONY
OASES
COLLECTIONS
BOOK STEPS
THE PEAK
STUDIOS
GROUP WORKING
OFFICES
MEETING AREA
MAKER SPACE
WORKING AREA
LIVING LAB

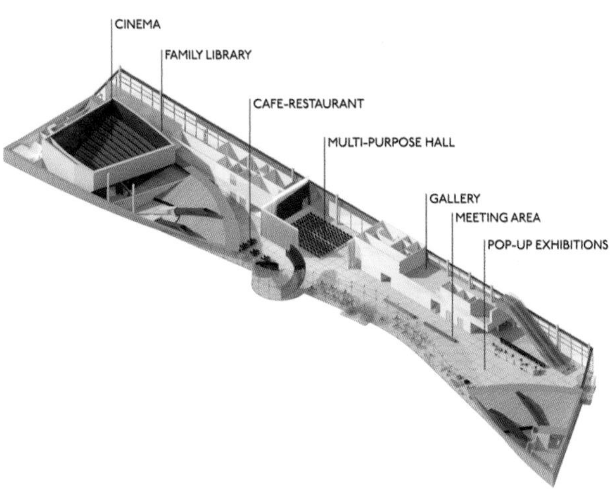
CINEMA
FAMILY LIBRARY
CAFE-RESTAURANT
MULTI-PURPOSE HALL
GALLERY
MEETING AREA
POP-UP EXHIBITIONS

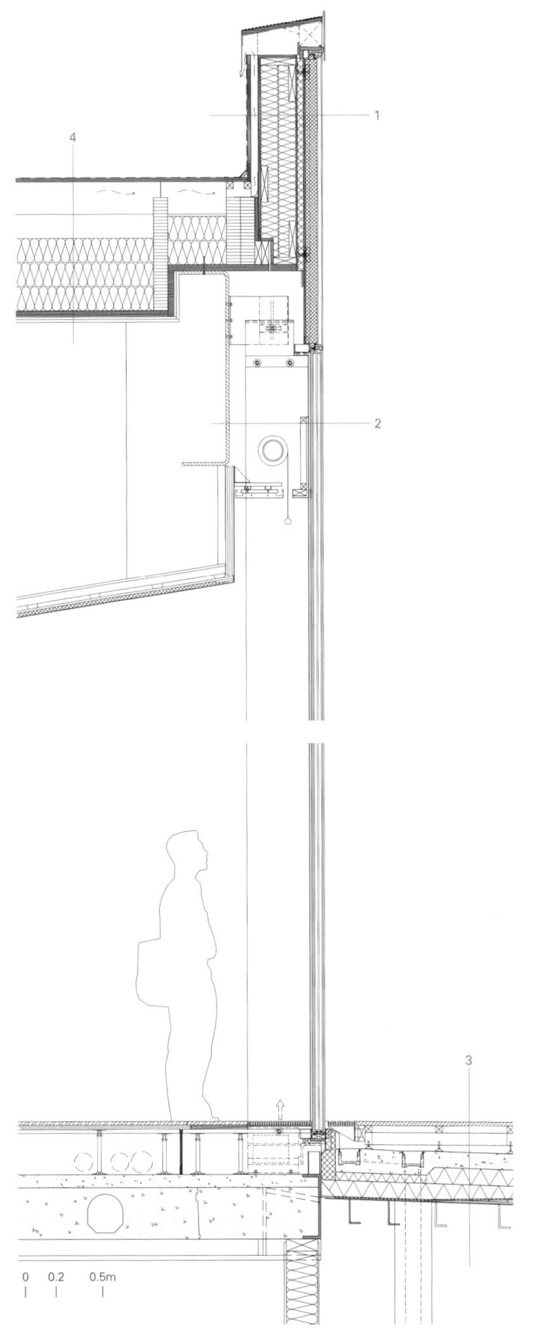

详图1 detail 1

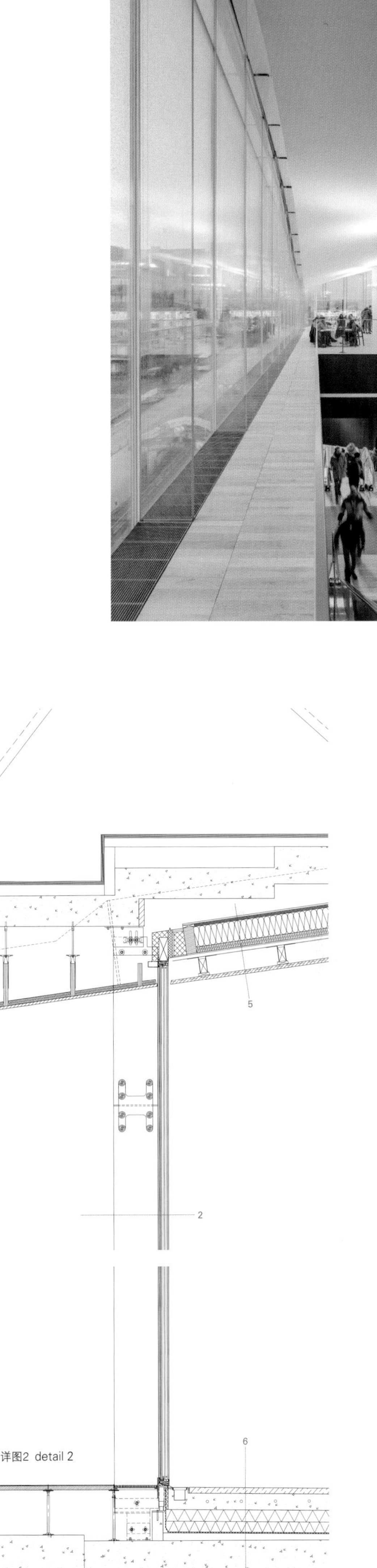

1. double glazed spandrel panel, digital printing to surface 1 and back painted, insulated unit bonded to rear side / mineral wool 50mm, fixing/movement zone, stainless steel fixings / timber parapet element 252mm / air gap 45mm / plywood 24mm / **underlay**: double bitumen membrane / elastomeric bitumen membrane with aluminum foil facing in white
2. **triple glazed insulated unit**: 88.4 laminated, tempered and digitally printed (surface 1) low-iron glass + 18mm ChrU spacer + 8mm tempered low-iron glass + 18mm ChrU spacer + 88.4 acoustic laminated, tempered and digitally printed (surface 9) low-iron glass / aluminum SG profile bonded to glass fin / glass fin, 4x10mm laminated and tempered low-iron glass
3. acetylated pine decking 26mm / pressure-impregnated joists 120mm / adjustable stainless steel support / concrete topping 100mm / separation sheet / extruded polystyrene insulation 3x100mm / drainage mat / waterproofing, bitumen / steel sheet 6mm with drainage falls / steel support structure
4. elastomeric bitumen membrane with aluminum foil facing in white / **underlay**: double bitumen membrane / **timber element**: plywood 24mm; cross-ventilated air gap, min. 100mm; mineral wool 450mm; vapour barrier; kerto-Q board; fire resistant gypsum panels 15+18mm; primary steel structure / **suspended ceiling**: thin gauge steel profiles radiused to suit ceiling geometry; monolithic acoustic system comprising of 40mm rock wool panel, filler and render
5. fire treated 33x88mm spruce board and lath 40x15mm, varnished / fire treated black painted counter lath 28x100mm / fire treated black painted counter batten 63x120mm, front face CNC cut to form curved geometry / painted steel support bracket of varying lengths / waterproofing steel sheet 0.6mm, galvanized and painted ventilation gap / insulated timber element 50/100mm granite paving / joints with sand-concrete mix / bedding concrete
6. 50mm / 120mm concrete slab with frost protection system / geotextile / 380mm XPS insulation / drainage matting / waterproofing membrane / reinforced concrete slab

详图2 detail 2

REGINA

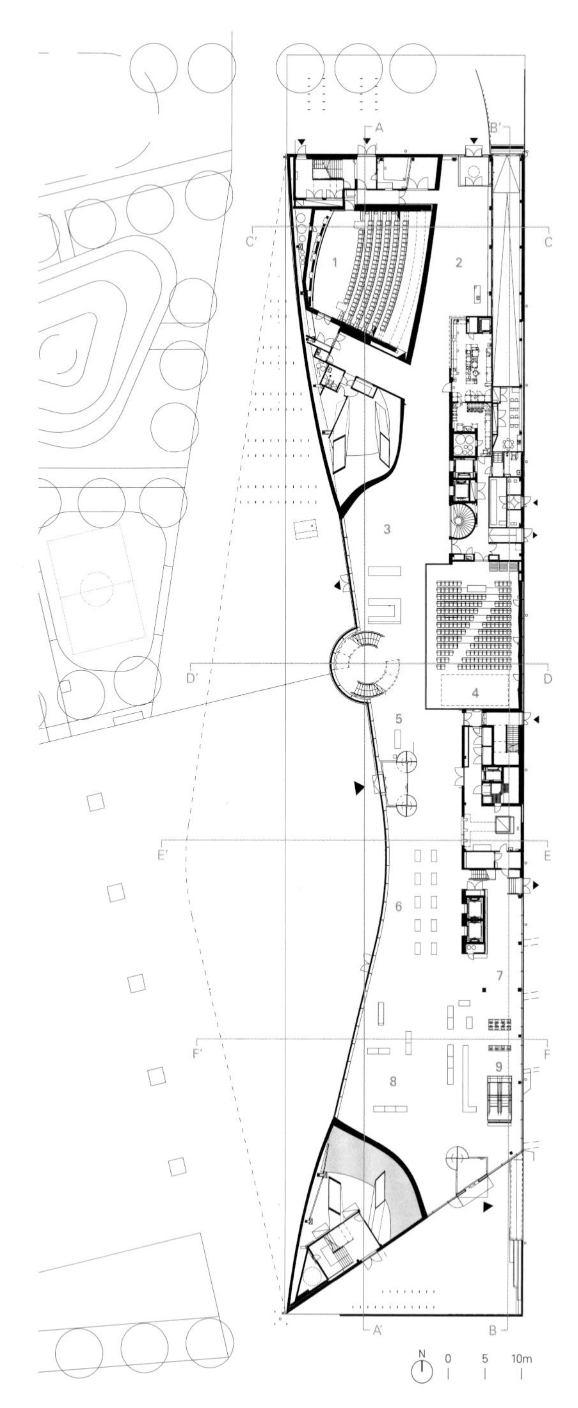

1. 电影院 2. 结账台 3. 餐厅 4. 多功能厅 5. 还书处 6. 门厅
7. 欧洲特快入口 8. 信息处 9. 用户终端
1. cinema 2. checkout 3. restaurant 4. multi-purpose hall 5. returns
6. foyer 7. europa experience 8. info 9. customer terminals
一层 ground floor

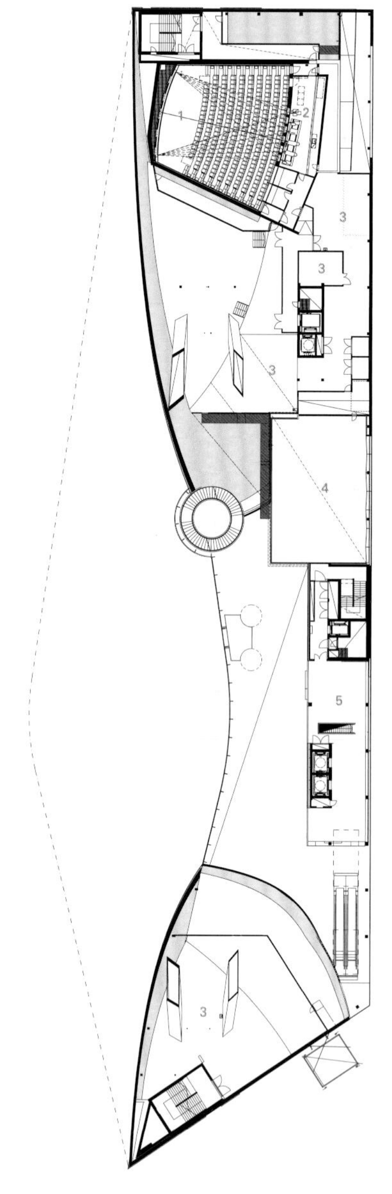

1. 电影院 2. 投影室 3. 设备空间 4. 多功能厅上部 5. 展览空间
1. cinema 2. projector room 3. technical space
4. multi-purpose hall top part 5. exhibition space
夹层 mezzanine floor

Institute's movie theater, is located on the ground floor along with a café restaurant whose seating spreads onto the square in summer. The middle floor, or "Attic", consists of flexible rooms arranged around intimate nooks and corners between the trusses of the bridge. Multi-function rooms – designed to accommodate both noisy and quiet activities – makers' spaces, and audiovisual recording studios are also provided. "Book Heaven", the top floor, is a vast open landscape topped with an undulating cloud-like white ceiling, punctured by circular skylights. The effect creates synergy between the modernist library and 21st Century technologies. The serene atmosphere invites visitors to read, learn, and think. From here, visitors can enjoy an unobstructed 360-degree panorama of the city center, or step out onto the terrace overlooking Kansalaistori square.
Maintenance and logistics spaces are located at basement level. These are kept to a minimum within publicly accessed levels, so as to create an open and accessible experience for users. In addition to the core operations, a wide range of auxiliary services provides public information on the city of

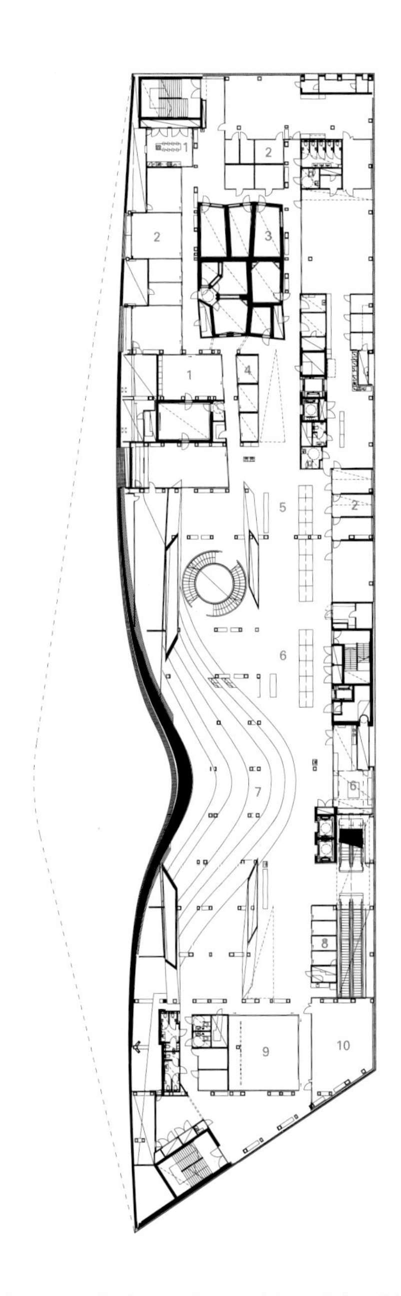

1. 学习空间 2. 团体活动室 3. 工作室 4. 游戏室 5. 工作站 6. 城市工作坊
7. 就坐台阶 8. 个人工作室 9. Kuutio活动空间 10. 阅览室
1. learning space 2. group room 3. studio 4. game room
5. workstation 6. urban workshops 7. sitting steps
8. individual working room 9. event space Kuutio 10. reading room
二层 first floor

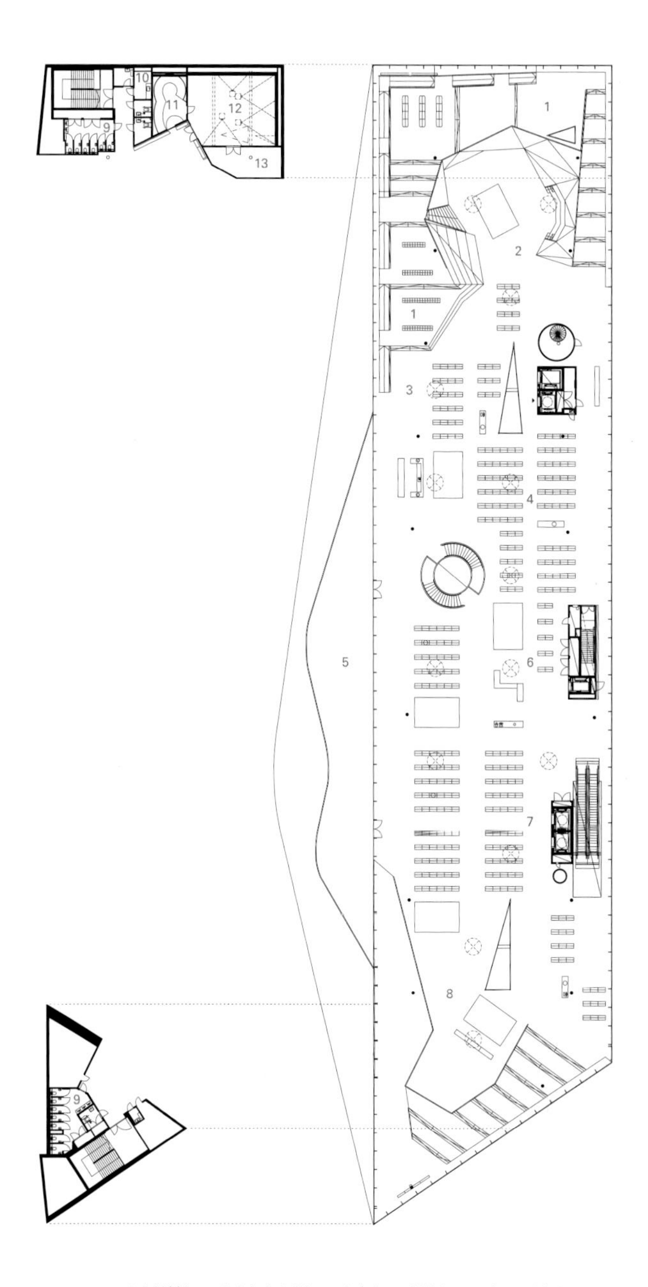

1. 儿童材料室 2. 儿童与家人房间 3. 杂志室 4. 小说室 5. 阳台 6. 预定室
7.非小说类文学作品室 8. 报纸室 9. 卫生间 10. 哺乳室 11. 童话室 12. 活动空间 13. 储藏室
1. children's materials 2. children and families 3. magazines 4. fiction
5. balcony 6. reservations 7. non-fiction 8. newspapers 9. wc
10. nursing room 11. fairytale room 12. event space 13. storage
三层 second floor

Helsinki and Europe. Other facilities include a community kitchen, an exhibition space, and children's playground both outside and inside the library.

Oodi has been built using local materials for local climate conditions. The facade is clad with 33mm thick Finnish spruce planks in a sweeping curve that extends the building outwards, creating a canopy above the square, blending the interior and exterior spaces, and creating shelter for public events. The complex curved geometry was designed and manufactured using precise algorithm-aided parametric 3D design methods. The building envelope uses the passive energy principles, making it a nearly Zero Energy Building (nZEB). Detailed analysis of the facade performance informed the environmental solutions and allowed the project team to minimize mechanical control systems. Copious daylight brightens the public areas through the glass facade, enhancing the quality of the library's interior as a multi-functional citizen's forum.

Bestsellerit
Bestsellerit

Bestsellerit
Kirjasto

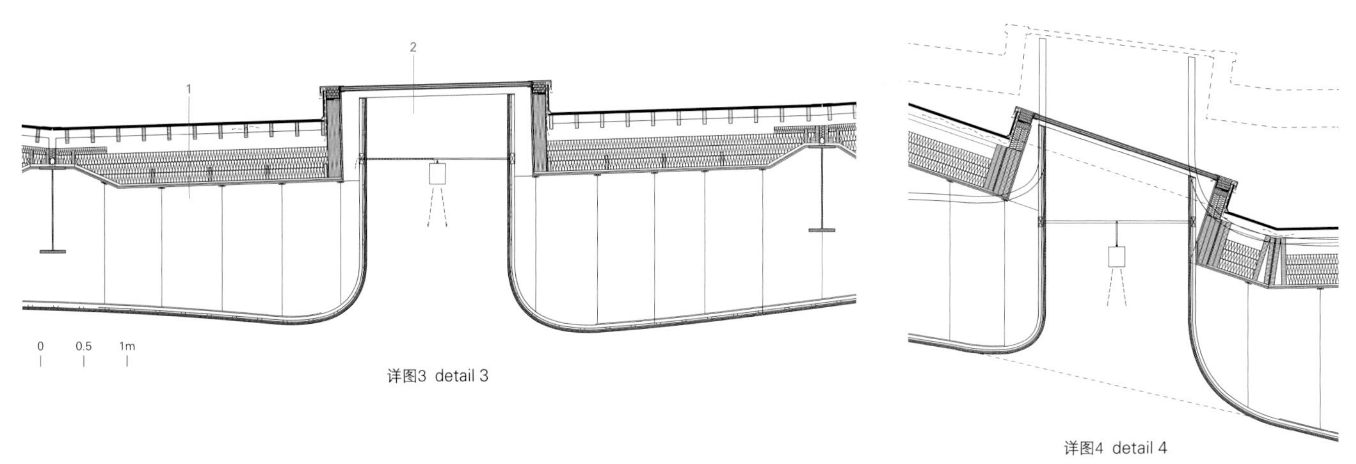

详图3 detail 3

详图4 detail 4

1. elastomeric bitumen membrane with aluminum foil facing in white / **underlay:** double bitumen membrane
timber element: plywood 24mm; cross-ventilated air gap, min. 100mm; mineral wool 450mm; vapor barrier; kerto-Q board; fire-resistant gypsum panels 15 + 18mm; primary steel structure
suspended ceiling: thin gauge steel profiles radiused to suit ceiling geometry; monolithic acoustic system comprising of 40mm rock wool panel, filler and render
2. skylight 3K glazing, 67mm / low-iron glass bonded to steel support structure / toughened and laminated glass

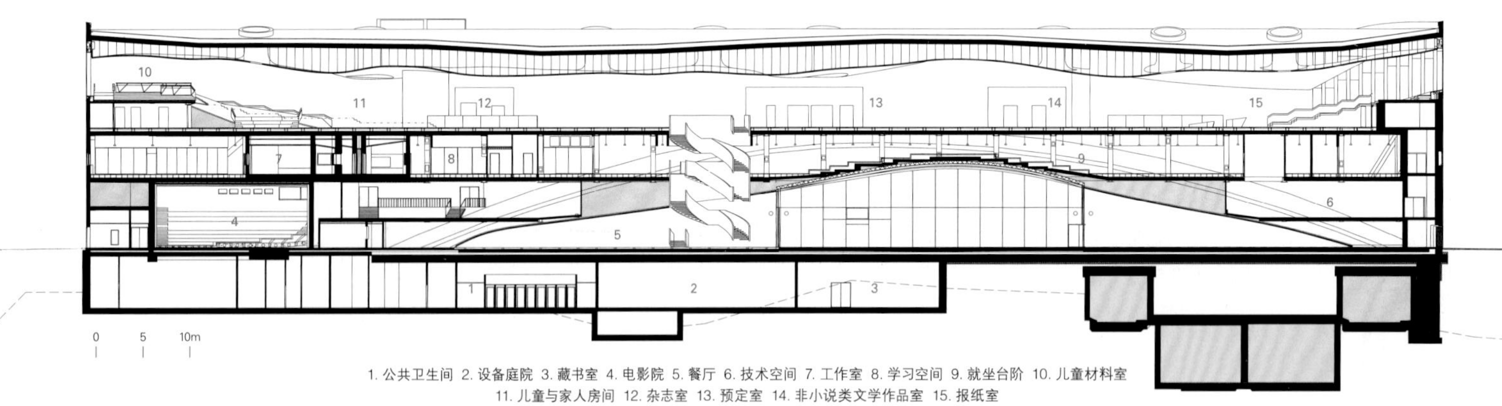

1. 公共卫生间 2. 设备庭院 3. 藏书室 4. 电影院 5. 餐厅 6. 技术空间 7. 工作室 8. 学习空间 9. 就坐台阶 10. 儿童材料室 11. 儿童与家人房间 12. 杂志室 13. 预定室 14. 非小说类文学作品室 15. 报纸室
1. public wc 2. service yard 3. book storage 4. cinema 5. restaurant 6. technical space 7. studio 8. learning space 9. sitting steps 10. children's materials 11. children and families 12. magazines 13. reservations 14. non-fiction 15. newpapers
A–A' 剖面图 section A-A'

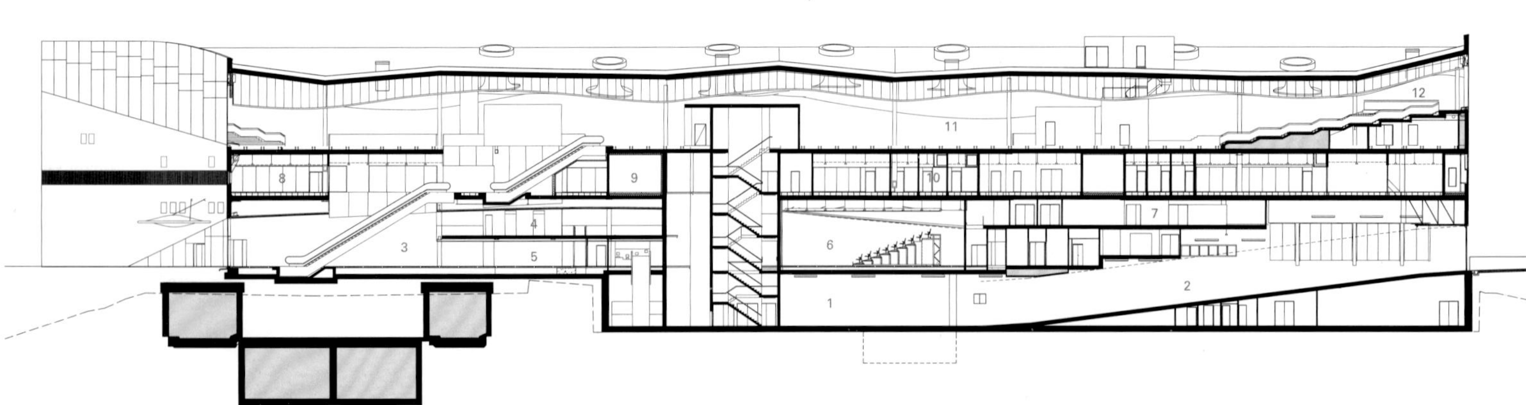

1. 设备庭院 2. 机动车坡道 3. 用户终端 4. 展览空间 5. 欧洲特快入口 6. 多功能厅 7. 技术空间 8. 阅览室 9. 城市工作坊 10. 团体活动室 11. 小说室 12. 儿童材料室
1. service yard 2. vehicular access ramp 3. customer terminals 4. exhibition space 5. europa experience 6. multi-purpose hall 7. technical space 8. reading room 9. urban workshops 10. group room 11. fiction 12. children's materials
B–B' 剖面图 section B-B'

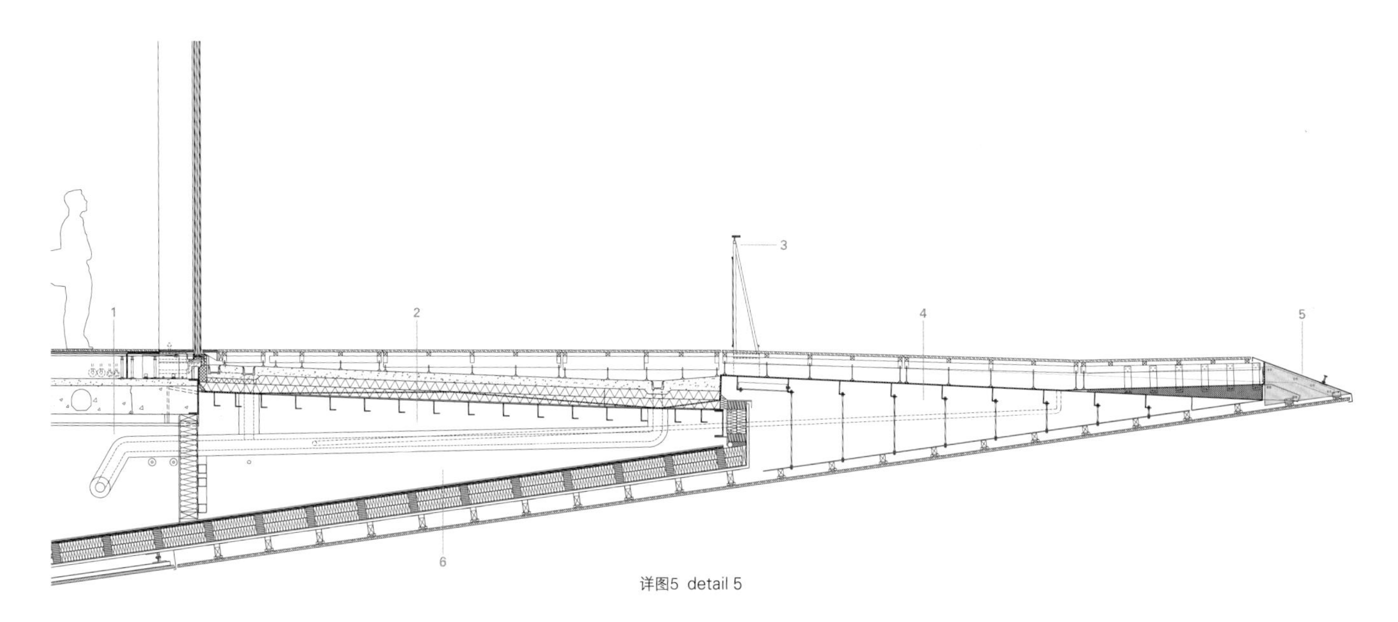

详图5 detail 5

1. oak boards 22mm / assembly floor system and technical cavity 300mm / concrete topping 80mm / hollow core slab 320mm 2. acetylated pine decking 26mm / pressure-impregnated joists 120mm / adjustable stainless steel support / concrete topping 100mm / separation sheet / extruded polystyrene insulation 3 x 100mm / drainage mat / waterproofing, bitumen / steel sheet 6mm with drainage falls / steel support structure 3. stainless steel railing with metal mesh fixed to structural steel sheet 4. acetylated pine decking 26mm / pressure-impregnated joists 120mm / adjustable stainless steel support / waterproofing, bitumen / steel sheet 6mm with drainage falls / steel support structure 5. balcony edge profile, 2mm stainless steel 6. fire treated 33 x 88mm spruce board and lath 40 x 15mm, varnished / fire treated black painted counter lath 28 x 100mm / fire treated black painted counter batten 63 x 120mm, front face CNC cut to form curved geometry / painted steel support bracket of varying lengths / waterproofing steel sheet 0.6mm, galvanized and painted / ventilation gap / insulated timber element

1. 技术空间 2. 结账台 3. 电影院 4. 团体活动室 5. 活动空间
6. 童话室 7. 卫生间 8. 儿童材料室
1. technical space 2. checkout 3. cinema 4. group room
5. event space 6. fairytale room 7. wc 8. children's materials
C-C' 剖面图 section C-C'

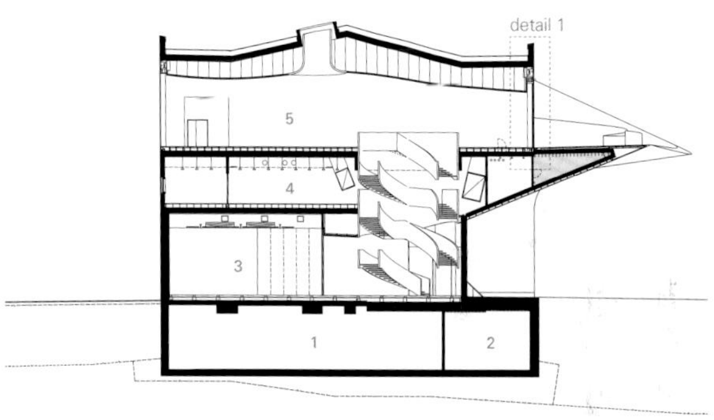

1. 设备庭院 2. 技术空间 3. 多功能厅 4. 城市工作坊 5. 小说室
1. service yard 2. technical space
3. multi-purpose hall 4. urban workshops 5. fiction
D-D' 剖面图 section D-D'

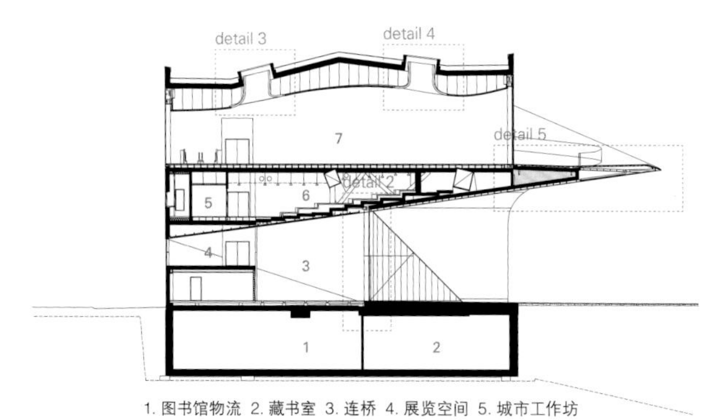

1. 图书馆物流 2. 藏书室 3. 连桥 4. 展览空间 5. 城市工作坊
6. 就坐台阶 7. 非小说类文学作品室
1. library logistics 2. book storage 3. bridge 4. exhibition space
5. urban workshops 6. sitting steps 7. non-fiction
E-E' 剖面图 section E-E'

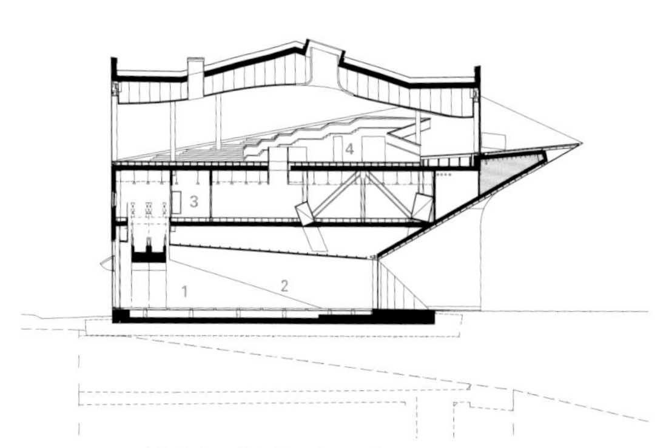
1. 客户终端 2. 信息处 3. 个人工作室 4. 报纸室
1. customer terminals 2. info
3. individual working room 4. newpapers
F-F' 剖面图 section F-F'

赫尔辛基对新图书馆的赞美
Helsinki's Ode to the New Library

Interview with ALA Architects

C3：Oodi中央图书馆在比赛结束的五年后正式开馆。图书馆项目的想法是如何成形的？

ALA建筑师事务所：新中央图书馆建设的倡议最初来自当时的文化部长Claes Andersson，他是在1998年提出的。2007年，赫尔辛基市的市长Jussi Pajunen启动了由独立顾问Mikko Leisti领导的中央图书馆审查程序。最终形成了题为《都市的中心——赫尔辛基的心脏》（莱斯蒂和马西奥，2008年）的报告，这为项目目标和愿景的实现都奠定了持久的基础。

2013年，当时的文化部长Paavo Arhinmäki宣布中央图书馆将是2017年芬兰纪念独立一百周年的一个重要项目。

C3：Oodi是什么意思？名字是如何选择的？

ALA建筑师事务所：Oodi这个词翻译成英语是"颂歌"的意思，一种以赞美一个事件或个人为特征的诗歌形式。

这个名字是在一次公开征名比赛之后选出的。由图书馆员、作家、公共咨询团队代表、城市和国家政策制定者和市场营销人员组成的评审团审议了1600个不同的名称。

赫尔辛基市副市长Ritva Viljanen代表评委们发言时说："这个名字能够表现出芬兰100周年纪念的象征意义。这个词很漂亮，这也是这个名字的一个优点。因为它是一个图书馆的名称，更应该与文学联系起来，从而与图书馆这个机构联系起来。

尽管有人对用图书馆的名字来铭记不同的作家提出了很好的论据，但是我们不想把容纳所有芬兰文学和文化的图书馆只献给对某一个人的记忆。在这些讨论的基础上，我们最后选择了'Oodi'作为中央图书馆的名字。"

C3：为什么ALA建筑师事务所能够赢得比赛？

ALA建筑师事务所：我们的设计是经过了包括初赛和复赛的国际设计竞赛选出的，这场比赛吸引了544名参赛者。所有参赛作品都是匿名提交的。评委们在关于入围作品的最终报告中，对他们的选择做出了如下评论：

"这个设计提案的质量非常高，执行时张弛有力，令人难忘。它为图书馆发展全新的功能理念提供了很好的前提。这座建筑具有独特的吸引力，也具备了成为赫尔辛基居民、图书馆用户以及图书馆员工的新标志性建筑的先决条件。"

C3：请解释一下为什么Oodi特别选址于此？

ALA建筑师事务所：Oodi图书馆占据赫尔辛基中部最后一个未经开发的场地。该地点特别适合于中央图书馆，因为它被可以代表现代自由社会的机构包围着——国家议会大厦、《赫尔辛基新闻报》报社、Musiikkitalo音乐厅和Kiasma现代艺术博物馆。

C3 Oodi Central Library finally opened its doors in five years after the competition. How did the library project come into being?
ALA Architects The initiative for the new Central Library building came originally from then Minister of Culture Claes Andersson in 1998. In 2007, the Mayor of Helsinki Jussi Pajunen launched a central library review process, led by the independent consultant Mikko Leisti. The final report of the review process, Heart of the Metropolis – the Heart of Helsinki, (Leisti and Marsio, 2008) laid a long-lasting foundation for goals and visions of the project.
In 2013 the Culture Minister at that time, Paavo Arhinmäki, declared that the Central Library would be a key project of Finland's commemoration of the centenary of its independence in 2017.

C3 What does Oodi mean and how was the name chosen?
ALA The word Oodi translates into English as "Ode", a poetic form marked by exaltation that praises an event or individual.
The name was chosen following a public competition in which members of the public were asked to suggest names. 1,600 different names were considered by a jury made up of librarians, authors, a representative of the public consultation group, city and national policymakers and marketers.
Speaking on behalf of the Judges, Helsinki Deputy Mayor Ritva Viljanen, said: "The name should support the symbolism of the Finland 100 centenary. It is also a merit to the name that the word is beautiful. Because it is a name of a library, it should be linked with literature and thereby with the library institution. We did not want to dedicate the library dealing with all of Finnish literature and culture to the memory of one person only, although good arguments were presented for library names dedicated to various authors. On these bases, we selected 'Oodi' as the name of the Central Library."

C3 Why did ALA Architects win the competition?
ALA Our design was chosen following a two-stage international competition which attracted 544 entries. All entries were anonymously submitted. In their final report on the shortlisted entries, the judges made the following comment on their selection:
"The proposal is of a very high quality, executed with relaxed broad strokes and memorable. It provides excellent premises for the development of a completely new functional concept for the library. The building has a unique appeal and the prerequisites to become the new symbolic building which Helsinki residents, library users, as well as the staff will readily adopt as their own."

C3 Please explain why this site was chosen for Oodi in particular.
ALA Oodi occupies one of the last undeveloped sites in Central Helsinki. The site is particularly appropriate for a central library, surrounded as it is by the institutions that typify a modern liberal society – the National Parliament Building, the offices of the Helsingin Sanomat Newspaper, The Musiikkitalo Concert Hall and Kiasma Modern Art Museum.

C3 What were the main challenges and constraints of the project?
ALA The main constraints of the project included a challenging site with long and thin proportions, crossed by the planning cor-

C3：该项目的主要挑战和限制是什么？

ALA建筑师事务所：项目的主要限制条件包括：具有挑战性的场地，这个场地是细长的，规划中的未来隧道的走廊将从这儿穿过；它位于许多主要市政机构之间，所有这些机构的设计都截然不同；图书馆的设计概要包括在一座建筑物内设置不同的功能和设施；最后，在规模如此之大的建筑物上首次使用木质立面。

C3：外部包层的木材是本地采购的吗？

ALA建筑师事务所：木立面覆盖着33mm厚的芬兰云杉木板，符合建筑向外延伸的连续曲线，在Kansalaistori广场上创造出了一个天篷，并将室内外空间融合在一起，为公共活动创造了一个有遮挡的区域。考虑到当地的气候条件，Oodi图书馆是使用当地材料建造而成的。

C3：有什么可以证明建筑在环保方面符合标准？

ALA建筑师事务所：图书馆在能源利用方面是极为高效的。它的能耗水平接近于零能耗建筑（nZEB）。这主要是由于建筑设备的高效率所决定的。建筑的材料效率得益于在外覆层中使用了木材。

在建筑外围护结构的设计方面，项目团队采用了被动式能源方法。对立面性能的详细分析得出了符合环境保护要求的解决方案，并使团队能够将对机械环境控制系统的要求降至最低，这有助于实现高度灵活的建筑解决方案。玻璃幕墙为公共区域提供了大量的功能性自然采光，减少了人造光的使用，提高了作为多功能市民广场的室内空间的质量。

精心设计的建筑设备的布局和优化使空间布局具有灵活性。所有的公共楼层均安装有活动地板，以便在建筑物预计150年的使用寿命期间进行更换。

C3：Oodi图书馆和芬兰的其他图书馆相比有什么不同？这家图书馆有何独特之处？

ALA建筑师事务所：Oodi让人们可以获得探索图书馆发展新理念这样一个让人兴奋的机会。Oodi的独特之处在于，图书馆内几乎所有的空间都是供公众使用的：例如，没有行政办公室、藏书处理中心或物流中心，这些功能在图书馆系统的其他区域产生。

Oodi图书馆关注用户的参与度，这使得它成为测试新思想和新式服务的完美场所，所有这些都将成为图书馆服务不断发展的一部分。Oodi图书馆将围绕持续的用户参与和互动的原则进行组织。它鼓励用户在空间内部自己创建各种活动，并将在图书馆员的支持下利用新设施开展活动。

Oodi图书馆还因为它包含新技术的方式而显得与众不同——让社会上每个人都能接触到图书馆的新技术。最新的制造技术，如激光切割机和3D打印机，将在创客空间和身临其境的3D虚拟现实空间（即The Cube）中免费使用，这些新技术将为探索新媒体体验提供机会。

ridor for a future tunnel; its location amongst a number of major civic institutions all with very different designs; the design brief for the library that included a multitude of functions and facilities within one building; and the timber facade that was used for the first time on a building of this scale.

C3 Is the timber for the exterior cladding locally sourced?
ALA The timber facade is clad with 33mm thick Finnish spruce planks that conform to the sweeping curve that extends the building outwards, creating a canopy above the Kansalaistori square and blending the interior and exterior spaces to create a shelter for public events. Oodi has been built using local materials with local climate conditions in mind.

C3 What are the environmental credentials of the building?
ALA The library will be extremely energy efficient. Its energy consumption level is that of a nearly Zero Energy Building (nZEB). This is mostly due to the efficiency of the building services. The material efficiency of the building has benefited among other things from the use of wood in the exterior cladding.
The design of the building envelope is intrinsic to the passive energy approach adopted by the project team. Detailed analysis of the facade performance has informed the environmental solutions and has allowed the team to minimize the requirement for mechanical environmental control systems, which facilitates the highly flexible architectural solution. The glass facades allow for large amounts of functional daylight in the public areas, reducing the use of artificial light and enhancing the quality of the interior spaces as a multi-functional citizens' forum.
The carefully planned placement and optimization of building services has enabled the flexibility of the spatial arrangements. All public levels are equipped with access floors to allow for changes in use during the estimated 150 year lifespan of the building.

C3 How does Oodi differ to other libraries in Finland? What's unique about this library?
ALA Oodi is an exciting opportunity to explore new ideas about what libraries might become. Oodi is unique in that practically all the space within the library is for public use: there are no administrative offices, collections handling or logistics centres, for example – these functions are handled elsewhere within the library system.
Oodi's focus on engagement with users makes it the perfect place to test new ideas and services as part of the constant evolution of library services. Oodi will be organized around the principle of constant user participation and interaction. Users are encouraged to create events within the space themselves and will have the support of librarians in setting up activities using the new facilities.
Oodi is also set apart by the way it embraces technology – placing new skills within the reach of everyone in society. The newest manufacturing technology, such as laser cutters and 3D printers will be free to use in the makerspace and an immersive 3D virtual reality zone, The Cube, will provide opportunities to explore new media experiences.

CALGARY
PUBLIC
LIBRARY

卡尔加里新中央图书馆
Calgary's New Central Library

Snøhetta + DIALOG

卡尔加里新中央图书馆终于面向公众开放了，扩建的目的是希望每年能够容纳之前两倍以上的游客来参观这个大约22300m²的扩建设施。作为卡尔加里自1988年奥运会以来最大的公共投资，它标志着这个快速扩张的城市进入了一个新的篇章。卡尔加里公共图书馆是北美最大的图书馆系统之一，超过一半的居民都是该图书馆的活跃的参与者。因此，新大楼为各类人群和承办各种类型的活动提供了不同的空间——从社会互动、学习到安静自省，面面俱到，支持着当今图书馆独特的城市功能。

该建筑处于复杂的城市环境之中，全面运转的轻轨交通线路以弯曲的半月形曲线的方式从地上到地下横穿场地而过，将市中心和奥运东村分隔开来。作为对地形的回应，该设计将主入口提升到了封闭的铁路线的上方。平缓的阶梯状斜坡一直延伸到建筑的中心，让来自四面八方的人们都可以与图书馆产生互动。室外的圆形剧场坐落在露台上，为室内功能向室外的扩展提供了场所。参照当地景观的植被设计将卡尔加里的山脉和草原引入到城市的景观中。以大门和桥梁的姿态展现在人们面前的广场入口重新建立起各区域之间的连接。

富有活力的三层玻璃立面由模块化的六边形格子图案组成。六边形上聚集起来的变化散布在由烧结玻璃和偶尔闪光的铝材交替组成的建筑曲面上。这些形状组合成一个个我们所熟悉的图案，如同一本打开的书、雪花似的线条，或者连锁的住宅，从而锚定了集体社区的想法。更重要的是，整个建筑体量被封闭在同一个模式中，使得所有的侧面又都可以作为建筑的"正面"呈现于世人面前。立面的水晶几何体被切掉，露出一个宽阔的木拱门。建筑入口的框架形式参考了该地区常见的奇努克云拱门。它的双曲壳完全由美国西部红雪松木板制成，这也是世界上最大的自由形式的木壳之一。它的有机形态和纹理又使得高大的建筑呈现出紧凑的形态。

沿着拱门进入到大厅和中庭，可以看到木质结构螺旋向上延伸了25m，使人们可以眺望到上面的天际。这个开放式中庭的四周边缘采用木板条包裹，在平面上像一个尖锐的椭圆形，向人们暗示了图书馆的交通流线和组织架构逻辑。梁和柱子所产生的韵律感让人想起古希腊建筑中的拱廊，这是一种公共的露天柱廊，用于聚会和学术交流。在内部，混凝土结构保持裸露，未进行任何修饰。材料色调的质朴旨在让人

感觉到图书馆是一个人人可以参与的地方，而不是一个神圣的藏书库。

六层高的图书馆为数字化、模拟操作、团体活动和个人互动分配了各种空间，从"娱乐"到"严肃"——低层的公共空间格外活跃，而到了高层都变成了安静的学习区域。在街道层面，多功能房间位于建筑的周边，增强了内部和外部的连通性。在一层，有一个儿童图书馆亦可作为儿童游戏室来使用，这里主要用于孩子们的手工艺和绘画活动、早期识字项目，以及全身室内游戏体验。最上层的大阅览室的设计犹如一个藏在珠宝盒中的空间，这是一个适合专注学习和唤醒灵感的空间。读者进入一个有着柔和光线和声音的过渡空间。在这里，垂直的木板条既提供了隐私，又提供了可视性，在没有使用实心墙的情况下，木板条的使用让自然光照亮了整个室内空间。

到达图书馆的最北端，你会发现这里有一个客厅，在这里可以俯瞰铁路线和两个街区的交汇点。这里充满了阳光，适合举办各种活动，这就使得建筑的"船头"本身不仅成为引导外部世界的灯塔，还将成为一个展望未来的地方——一个重新激活卡尔加里文化、学习和社区精神的理想所在。

Calgary's New Central Library opened its doors to the public with an aim to welcome over twice as many annual visitors to approximately 22,300m² of expanded facilities. As Calgary's largest public investment since the 1988 Olympics, it signals a new chapter of the rapidly expanding city. Calgary Public Library is one of the largest library systems in North America, where more than half of its residents are active cardholders, and accordingly, the new building provides spaces for all types of people and activities – from social interaction, learning, to quiet introspection, championing the unique civic function of libraries today.

The building is sited within a complex urban condition, where a fully operational Light Rail Transit Line crosses the site from above to below ground on a curved half-moon

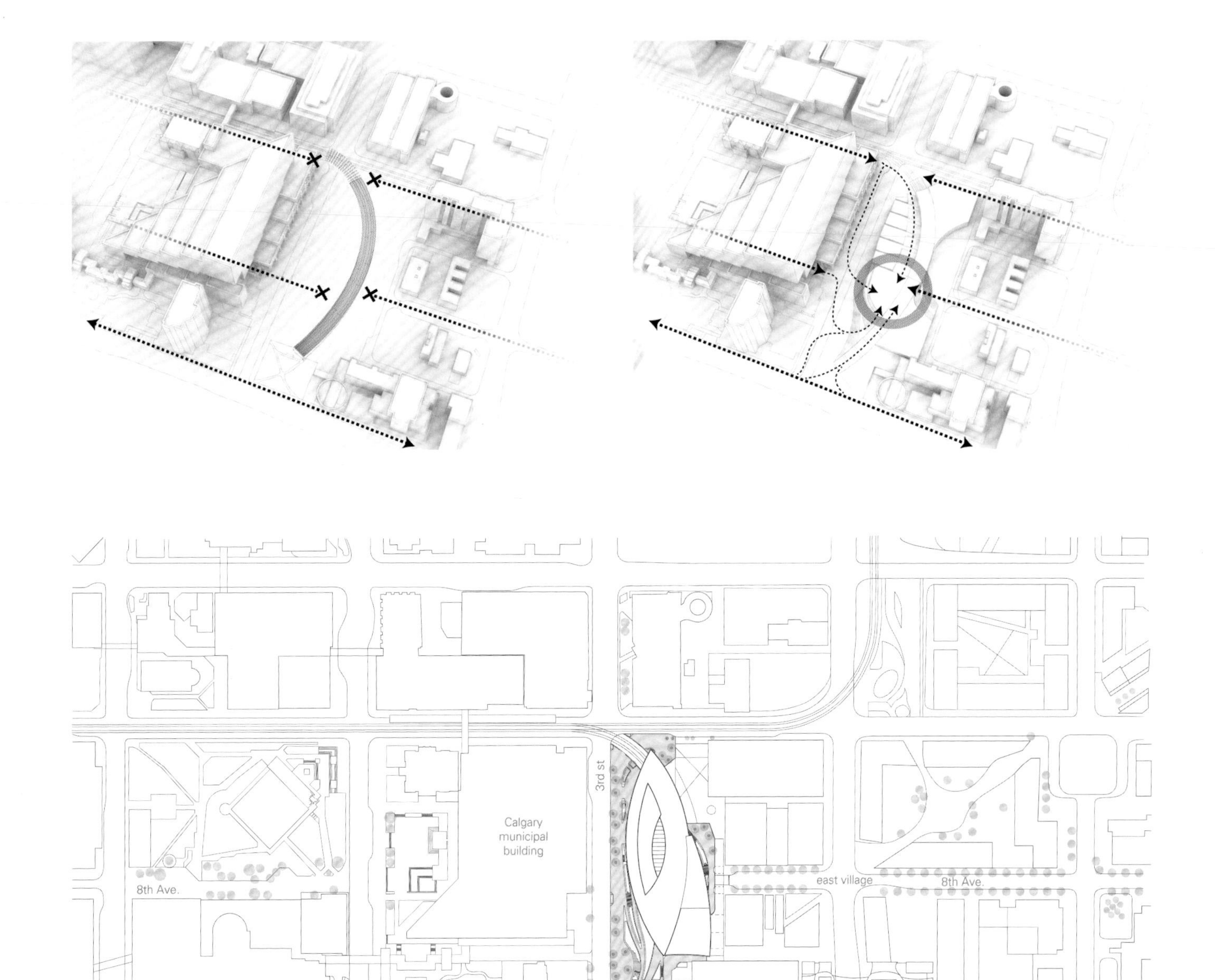

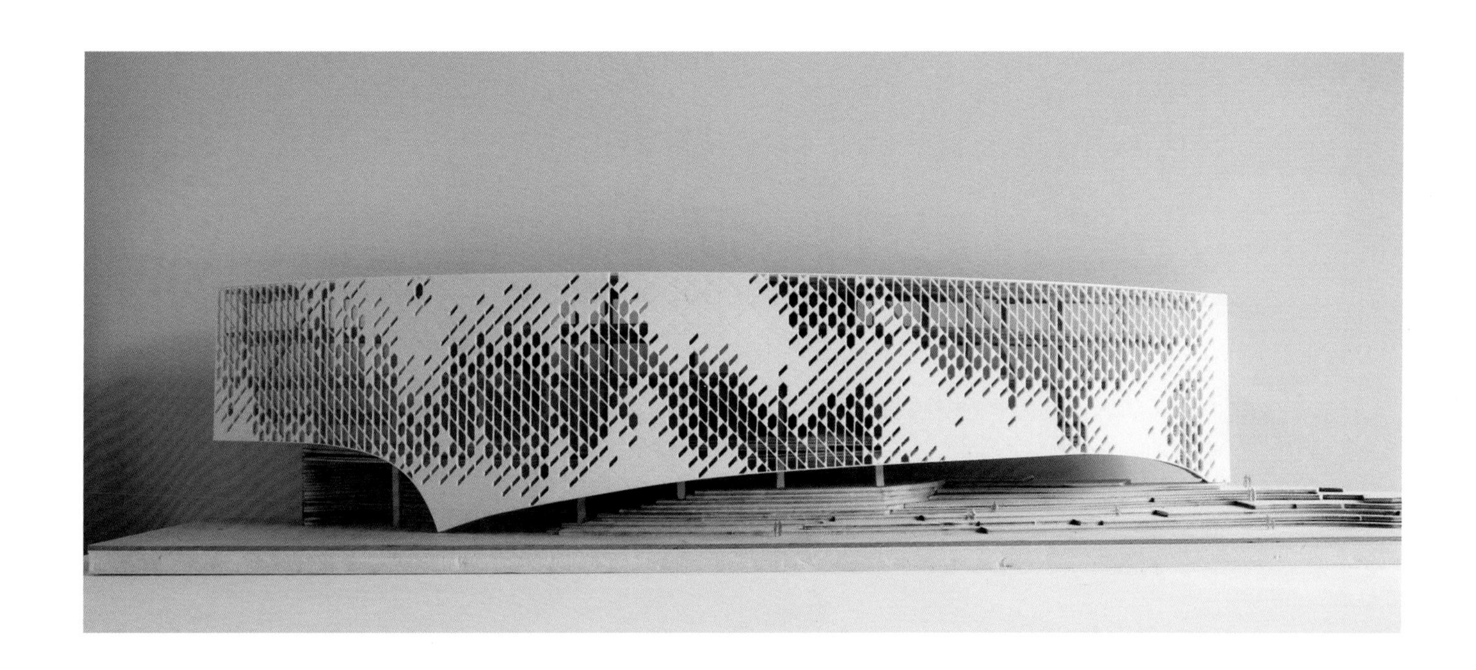

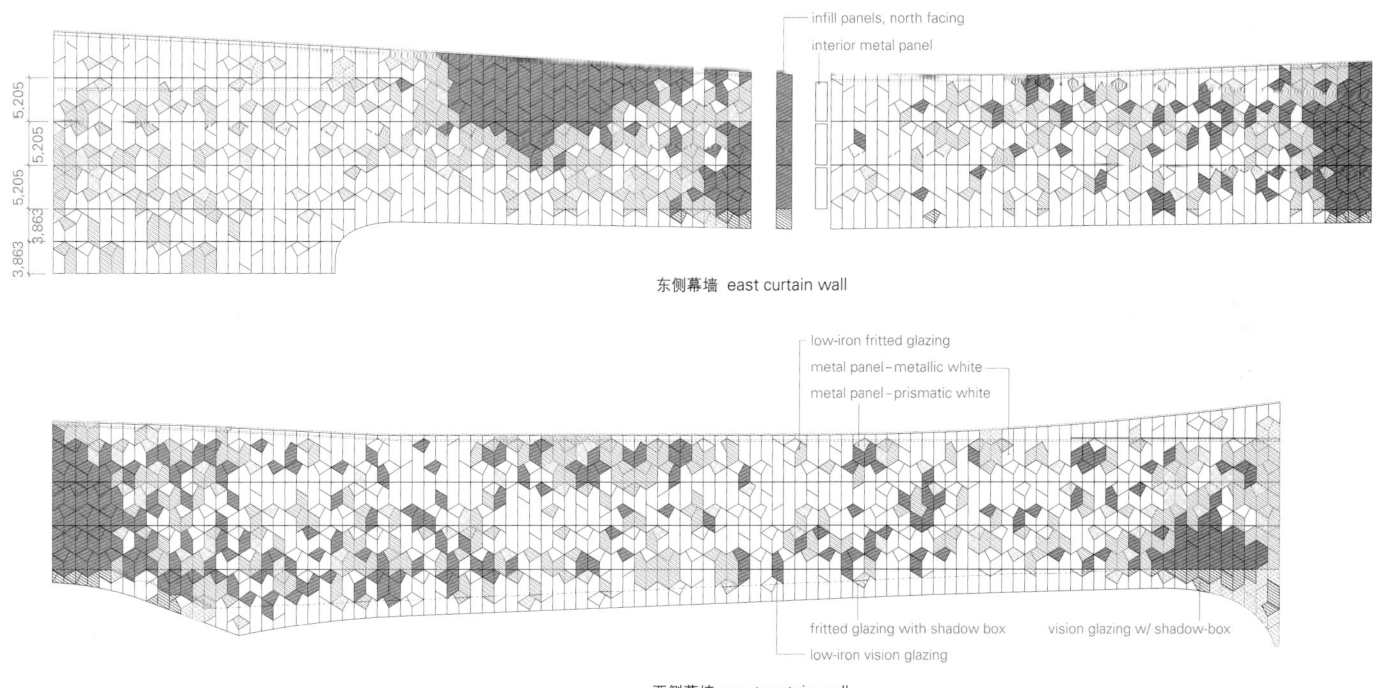

东侧幕墙 east curtain wall

西侧幕墙 west curtain wall

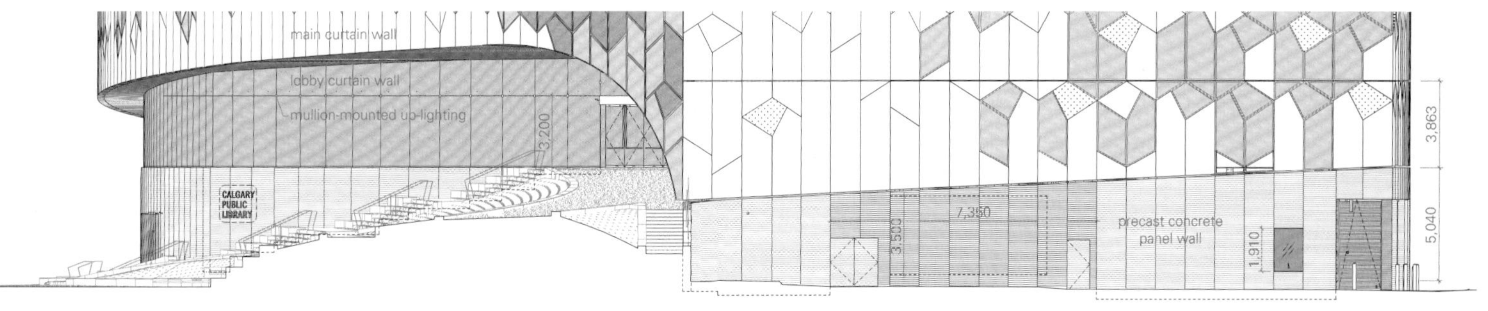

南立面——入口详图 south elevation _ entrance detail

CREATESPACE

path, dividing Downtown and East Village. In response, the design lifts the main entry over the encapsulated train line. Gently terraced slopes rise up to the heart of the building, allowing people to arrive from every direction to interact with the library. Outdoor amphitheaters nestled into the terraces provide places for inner programs to spill outside. Plantings that reference the native landscape draw Calgary's mountains and prairies into the cityscape. As a portal and a bridge, the entry plaza reestablishes a connection across the site.

The dynamic, triple-glazed facade is composed in modular hexagonal pattern. Aggregated variations on the hexagon form scatter across the building's curved surface in alternating panels of fritted glass and occasional iridescent aluminum. These shapes emerge as familiar forms such as an open book, snowflake-like linework, or interlocking houses, anchoring the ideas of the collective community. More importantly, the entire volume is enclosed in the same pattern, allowing all sides to function as the "front" of the building. The crystalline geometry of the facade is carved away to reveal an expansive wood archway. Framing the entrance of

项目名称：Calgary's New Central Library
地点：Calgary, Alberta, Canada
建筑师：Snøhetta + DIALOG
设计，景观，室内，家具、固定装置和设备，打磨、标识&导向系统：Snøhetta
执行建筑师、景观设计师：DIALOG
结构工程师：Entuitive
机械工程师：DIALOG
电气&照明工程师：SMP Engineering
IT&视听技术：McSquared System Design Group, Inc.
客户：Calgary Municipal Land Corporation (CMLC)
总建筑面积：22,293.73m^2 (2/3 larger than original library)
造价：CAD 245 million (total project cost), CAD 147 million (hard cost)
竣工时间：2018
摄影师：©Michael Grimm (courtesy of the architect)

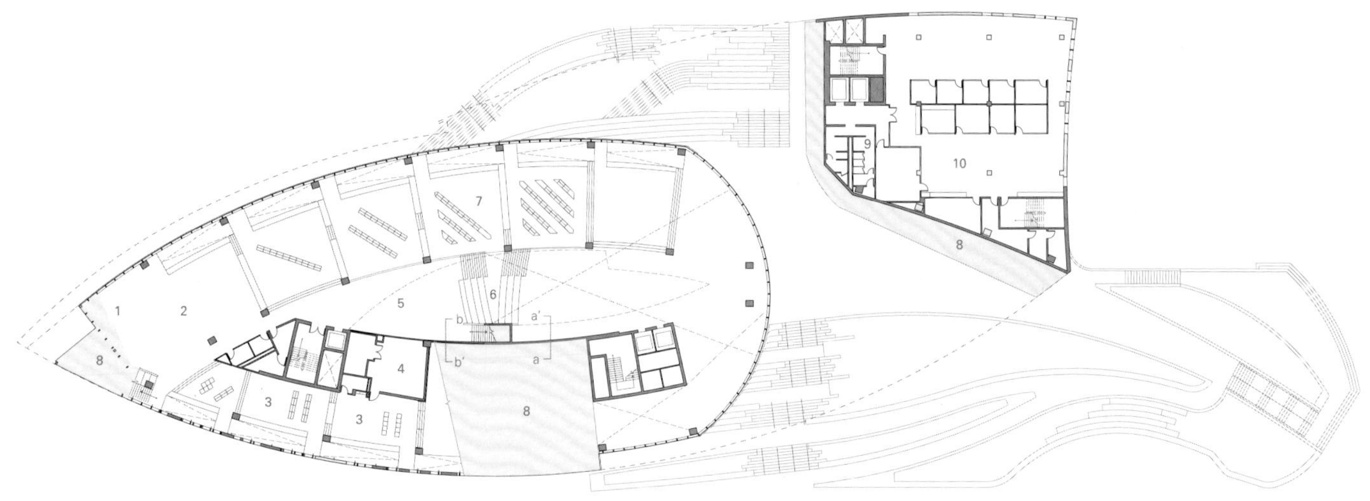

1. 北侧俯瞰区 2. 咖啡厅 3. 儿童图书馆 4. 早教中心&手工 5. 中庭阅览室 6. 大楼梯 7. 主题阅读&藏书室
8. 上空空间 9. 卫生间 10. 图书馆运营&培训
1. north overlook 2. cafe 3. children's library 4. early learning center & crafts 5. atrium reading lounge
6. grand stair 7. themed reading & collections 8. void 9. WC 10. library operations & training
夹层 mezzanine

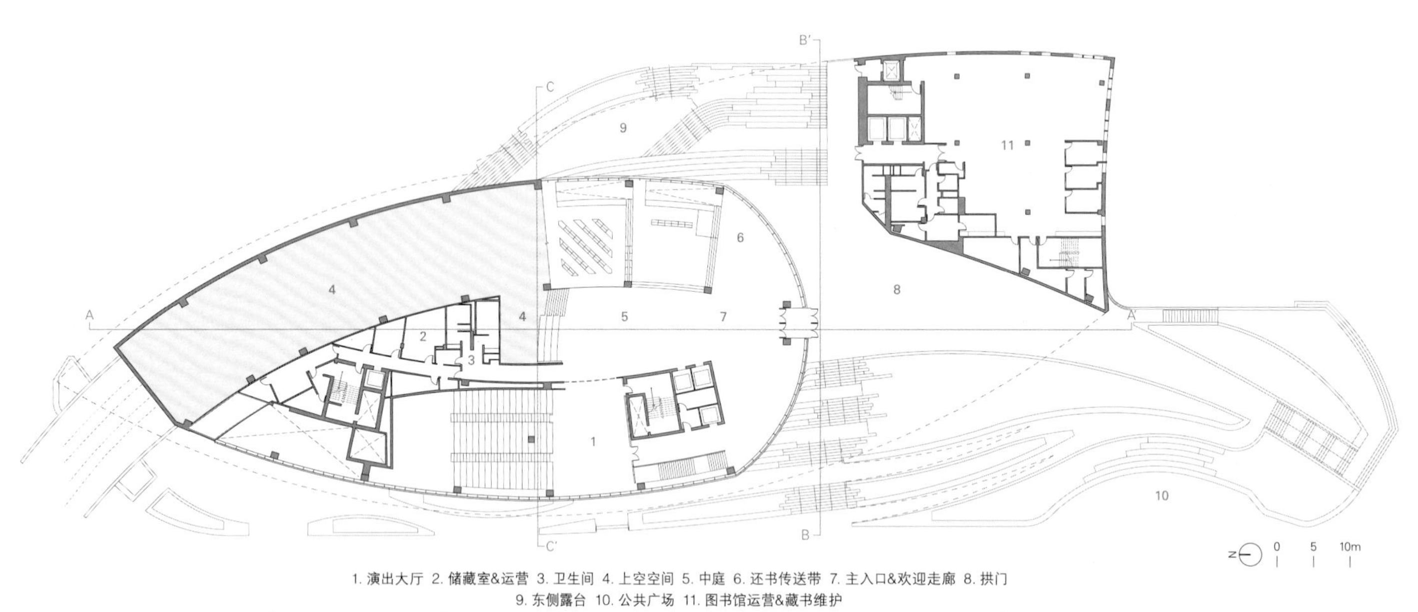

1. 演出大厅 2. 储藏室&运营 3. 卫生间 4. 上空空间 5. 中庭 6. 还书传送带 7. 主入口&欢迎走廊 8. 拱门
9. 东侧露台 10. 公共广场 11. 图书馆运营&藏书维护
1. performance hall 2. storage & operations 3. WC 4. void 5. atrium 6. book return conveyor 7. main entrance & welcome gallery
8. archway 9. east terrace 10. public plaza 11. library operations & collections maintenance
一层 first floor

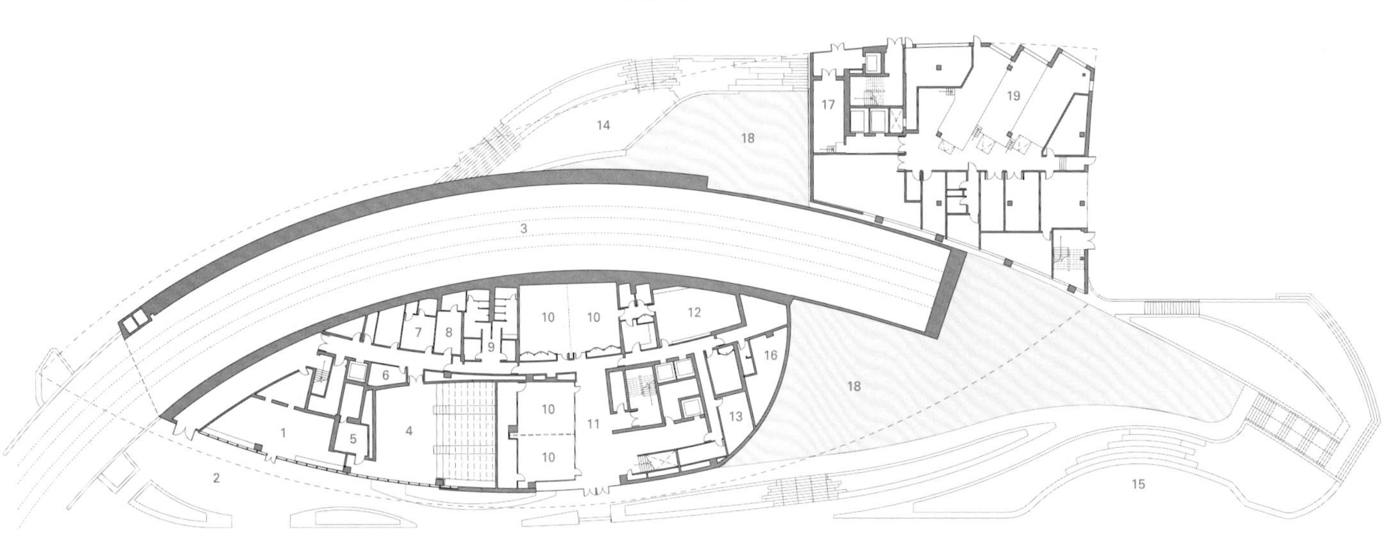

1. 咖啡厅 2. 室外座位 3. 轻轨交通 4. 演出大厅 5. 后台 6. 控制室 7. 温室 8. 更衣室 9. 卫生间 10. 多功能室 11. 社区大厅
12. 会议室 13. 通信室 14. 东侧露台 15. 公共广场 16. 水池 17. 自行车停车处 18. 上空空间 19. 卸货区
1. cafe 2. outdoor seating 3. light rail transit 4. performance hall 5. backstage 6. control room 7. green room
8. dressing room 9. WC 10. multipurpose room 11. community lobby 12. boardroom 13. communications room
14. east terrace 15. public plaza 16. cistern 17. bicycle storage 18. void 19. loading
地下一层 first floor below ground

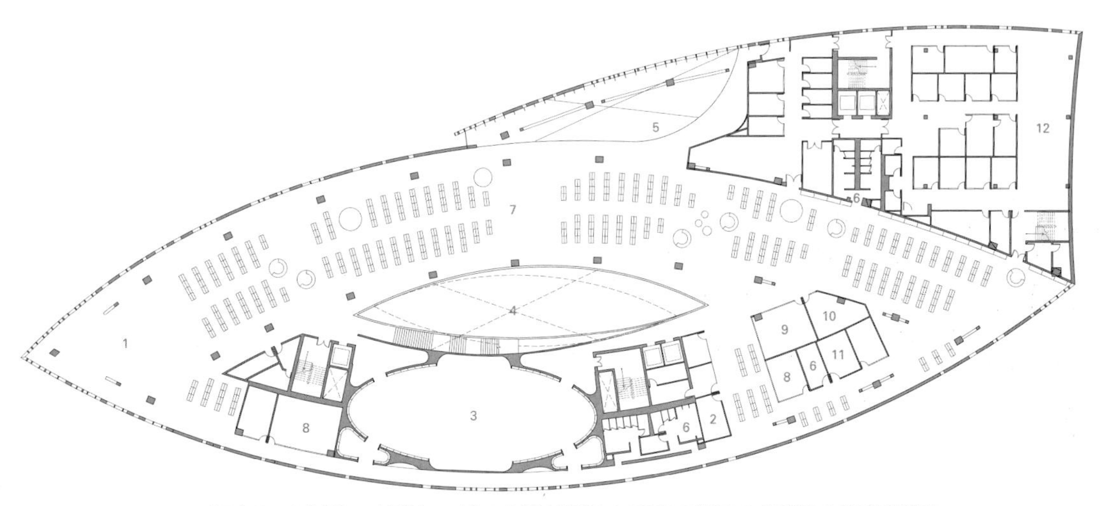

1. 社区客厅 2. 宗教房间 3. 大阅览室 4. 中庭 5. 东侧乡村俯瞰区 6. 卫生间 7. 藏书室 8. 多功能室 9. 手工艺术工作室 10. 年长者指导区 11. 故事工作室&媒体实验室 12. 图书馆运营

1. community living room 2. interfaith room 3. great reading room 4. atrium 5. east village overlook 6. WC 7. collections 8. multipurpose room 9. craft artist studio 10. elder's guidance circle 11. story studio&media lab 12. library operations

四层 fourth floor

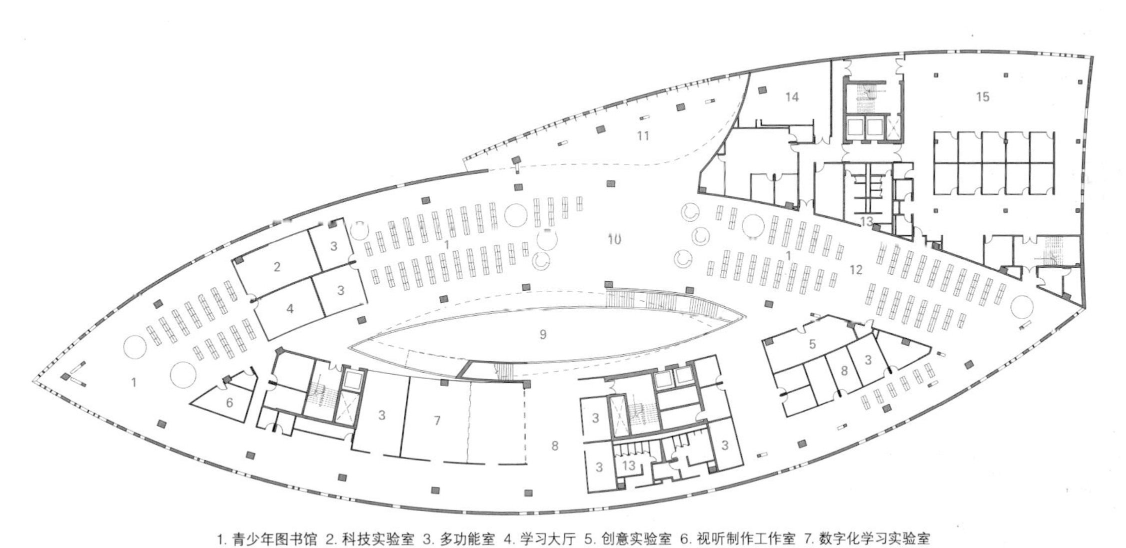

1. 青少年图书馆 2. 科技实验室 3. 多功能室 4. 学习大厅 5. 创意实验室 6. 视听制作工作室 7. 数字化学习实验室 8. 数字化公共空间 9. 中庭 10. 藏书室 11. 东侧阅览室 12. 新顾客服务处 13. 卫生间 14. 员工休息室 15. 图书馆运营

1. teen's library 2. technology lab 3. multipurpose room 4. study hall 5. idea lab 6. audio/visual production studios 7. digital learning lab 8. digital commons 9. atrium 10. collections 11. east reading room 12. services for newcomers 13. WC 14. staff lounge 15. library operations

三层 third floor

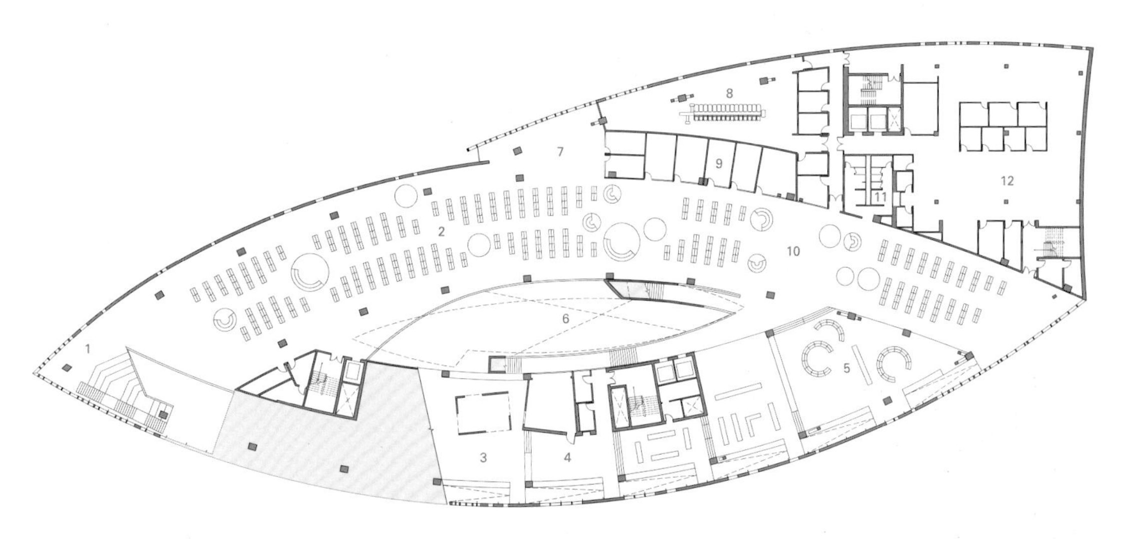

1. 北侧俯瞰区 2. 藏书室 3. 早教中心 4. 幼童区 5. 问答区 6. 中庭 7. 作者专区 8. 图书分类室 9. 会议&多功能室 10. 特殊服务处 11. 卫生间 12. 图书馆运营

1. north overlook 2. collections 3. early learning center 4. toddler's nook 5. questionarium 6. atrium 7. writer's nook 8. book sorting room 9. meeting&multipurpose rooms 10. special services 11. WC 12. library operations

二层 second floor

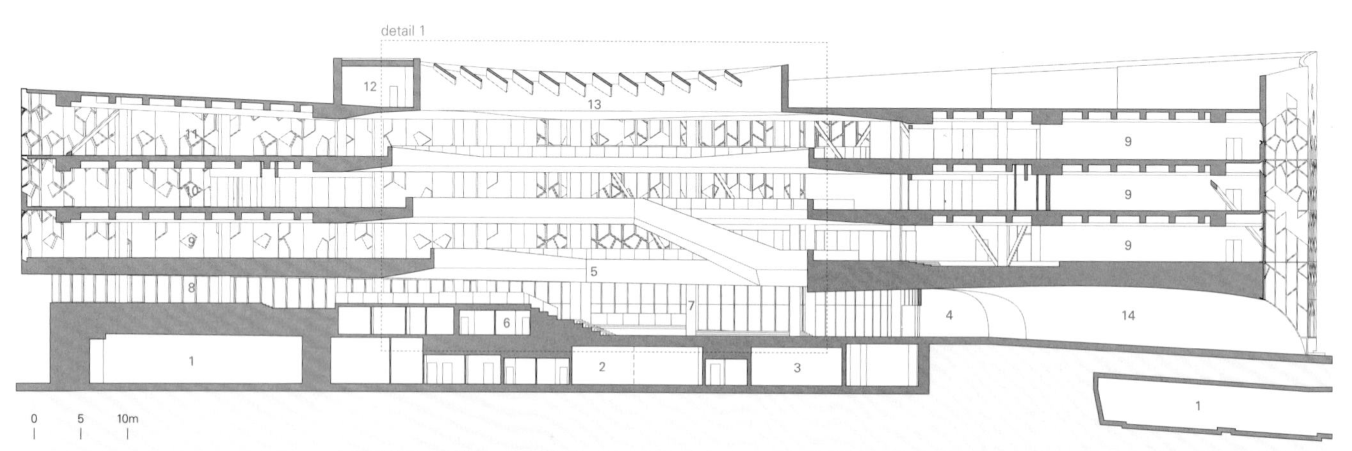

1. 轻轨交通 2. 多功能室 3. 会议室 4. 主入口 5. 中庭 6. 卫生间 7. 阅览室&主题藏书室 8. 咖啡厅 9. 藏书室 10. 少年图书馆 11. 客厅 12. 机械设备间 13. 采光天窗 14. 拱门

1. light rail transit 2. multipurpose room 3. boardroom 4. main entrance 5. atrium 6. WC 7. reading & themed collections 8. cafe 9. collections 10. teen's library 11. living room 12. mechanical 13. skylight 14. archway

A-A' 剖面图 section A-A'

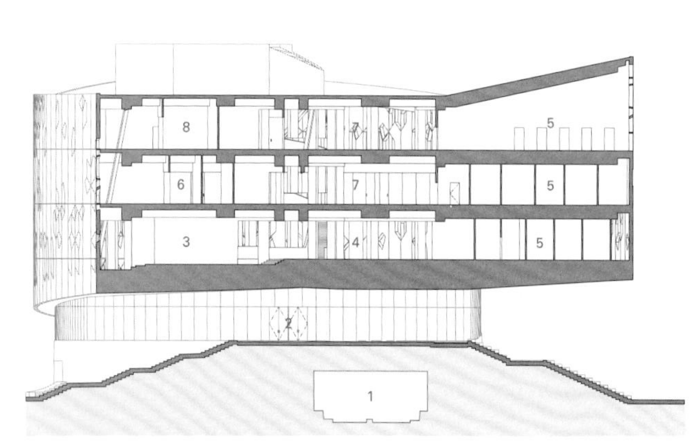

1. 轻轨交通 2. 主入口 3. 媒体收藏室 4. 学习区 5. 办公室 6. 多功能室 7. 藏书室 8. 工作室

1. light rail transit 2. main entrance 3. media collection 4. study area 5. offices 6. multipurpose room 7. collections 8. studio

B-B' 剖面图 section B-B'

1. 轻轨交通 2. 多功能室 3. 演出大厅 4. 中庭 5. 作者专区 6. 藏书室 7. 功能房 8. 游戏区 9. 东阅览室 10. 计算机房 11. 大阅览室 12. 机械设备间 13. 采光天窗 14. 东侧露台

1. light rail transit 2. multipurpose room 3. performance hall 4. atrium 5. writer's nook 6. collections 7. program room 8. playground area 9. east reading room 10. computer lab 11. great reading room 12. mechanical 13. skylight 14. east terrace

C-C' 剖面图 section C-C'

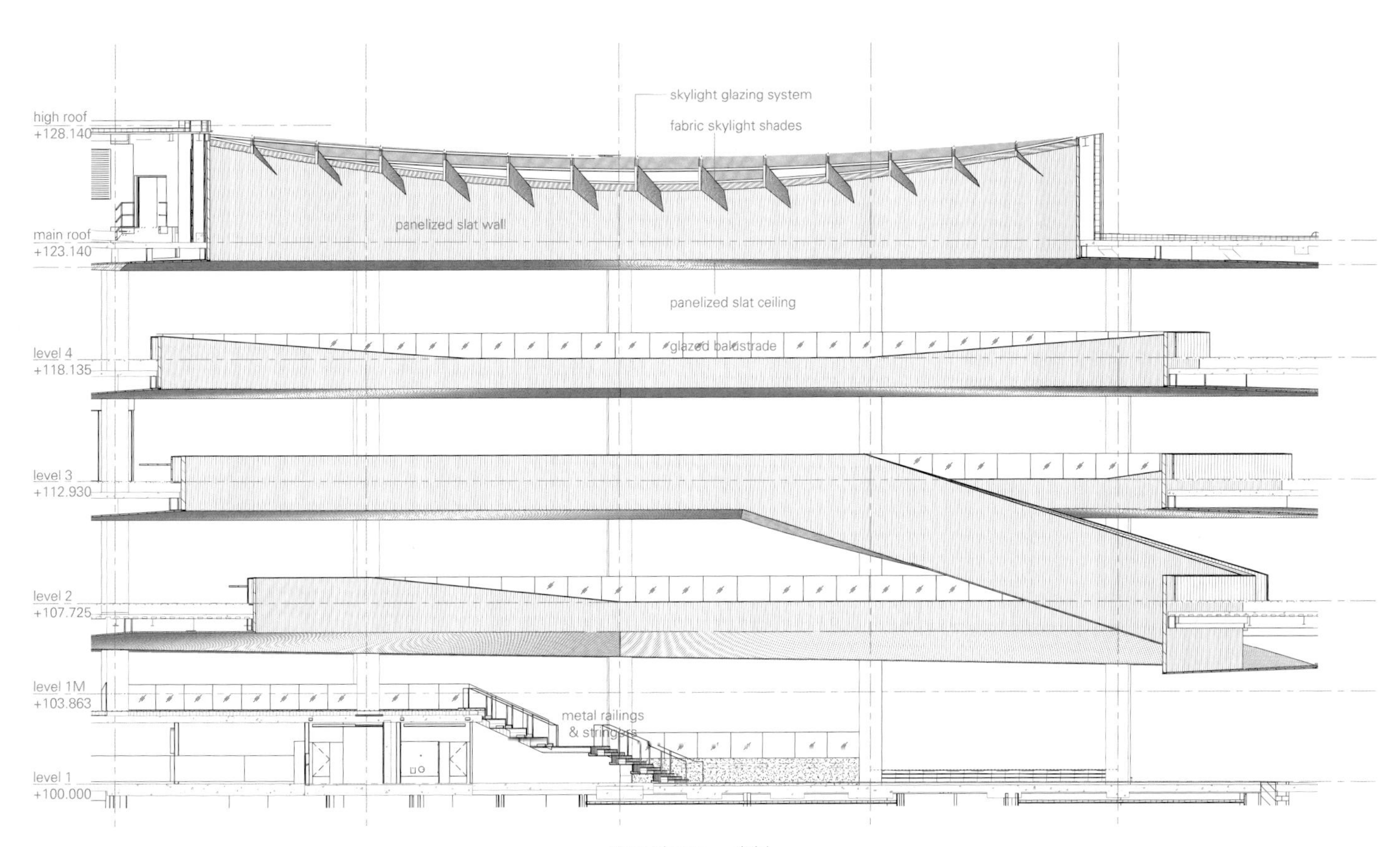

详图1（东立面——中庭）
detail 1 (east elevation_atrium)

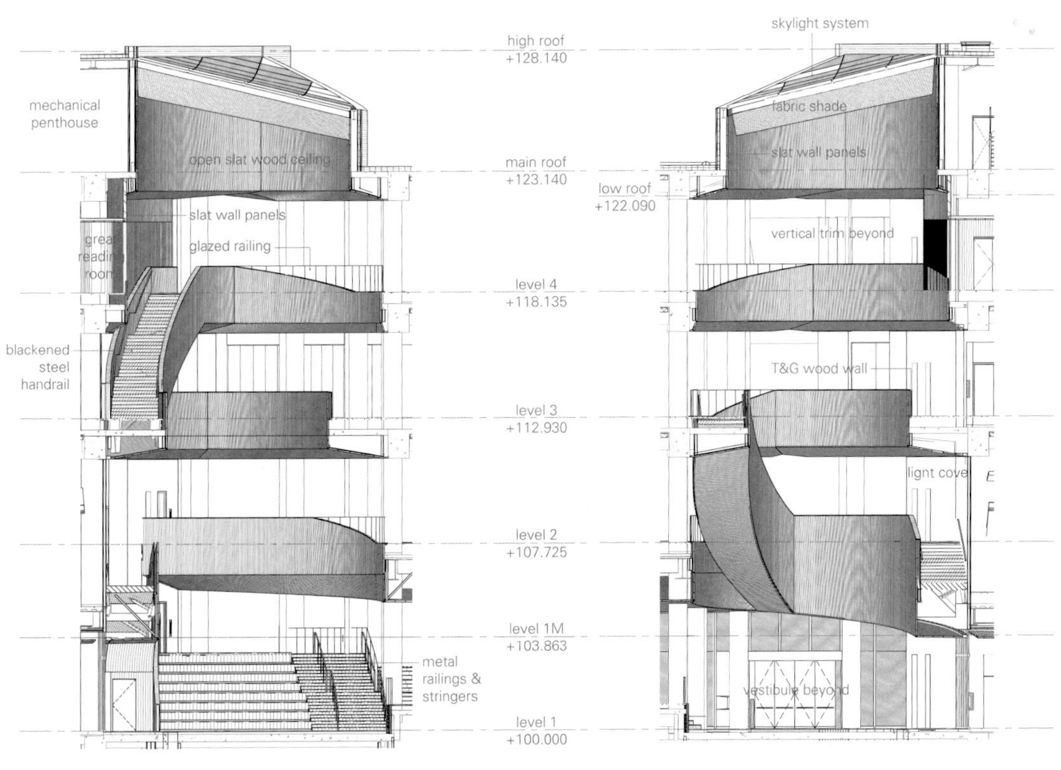

a–a' 剖面详图（北立面——中庭）
detail a-a' (north elevation_atrium)

b–b' 剖面详图（南立面——中庭）
detail b-b' (south elevation_atrium)

Children's Library

the building, the form references the Chinook cloud arches that are common in the region. Made entirely of western red cedar planks, its double-curved shell is one of the largest freeform timber shells in the world. Its organic form and texture bring the large building down to an intimate scale. As the archway continues into the lobby and atrium, the wood spirals upwards over 25m to a view of the sky through the oculus. Wood slats line the perimeter of the open atrium, shaped like a pointed ellipse in plan, hinting people the circulation and organizational logic of the library. The rhythm of beams and columns are reminiscent of a stoa, the public open-air colonnades of ancient Greek architecture for gathering and intellectual exchange. Inside, the concrete structure is left exposed and unfinished. The rawness of the material palette is intended to give a sense that the library is a place of engagement, rather than a sacrosanct repository for books.

The six-story library distributes a variety of spaces for digital, analog, group, and individual interactions in a spectrum of "fun" to "serious" - lively public activities on the lower levels and quieter study areas on the upper levels. At the street level, multi-purpose rooms are located along the perimeter of the building, enhancing the connectivity between inside and outside. On the ground floor, a children's library offers playhouses for crafts and drawing-based activities, early literacy programs, and a full-body indoor play experience. The great reading room at the uppermost level, conceived as a tucked in jewel box, is a space for focused study and inspiration. Readers enter through a transitional space with softened light and acoustics. Within, vertical wood slats provide both privacy and visibility, defining an interior space without using solid walls, illuminated with natural light through the slats.

Arriving at the northernmost point of the library, one finds the living room, overlooking the train line and the meeting point of the two neighborhoods. Filled with light and activities, this prow of the building will not only serve as a beacon to those outside, but also as a prospect for looking back out – a fitting vantage point that re-energizes the spirit of culture, learning, and community in Calgary.

回念：建筑嵌入、连续性和变化

Palimpsest: Architecture Continuity a

在历史和遗产的背景下，针对新建筑所进行的建筑嵌入的过程是复杂的，它需要从美学、功能性、技术性和公共观点等方面进行深入的推理和论证。

要考虑的最基本的方面包括建筑的位置、视觉和空间质量、建筑在其周围环境中的构成方式以及建筑融入城市历史环境的方式。

我们可以通过使用新的外层来改造一栋建筑，并在新建建筑与现有的结构之间建立一种对话。嵌入部分应该与原有的建筑有所区别，但要考虑到两者的和谐性。新建筑的目标应该是与传统建筑达成一种共识，而不是寻求压倒传统建筑。有一种方法就是当成复合的艺术作品来处理，这样操作会在过去的痕迹和现在的存在之间实现微妙的互动，同时保持空间的最基本的和谐。

The architectural insertion of new building layers in a historical and heritage context is complex and requires in-depth reasoning and justification from an aesthetic, functional, technical and public point of view.
Among the most fundamental aspects to consider are the locality, visual and spatial quality of architecture, the way a building is composed within its surrounding, and the manner of its integration into a historic urban context.
Adopting a new layer transforms a building and establishes a dialogue with the existing fabric. Insertions should be differentiated but harmonious. The goal should be that new construction achieves a consensus with a traditional setting, rather than seeking to overpower it. One is dealing with composite work of art that allows subtle interplay between past and present while preserving the essential harmony of a space.

大馆传统与文化中心_Tai Kwun, Center for Heritage & Arts/Herzog & de Meuron
格拉纳达费德里科·加西亚·洛尔卡中心_Federico García Lorca Center in Granada/MX_SI(Mendoza Partida + BAX studio)
卡托维兹西里西亚大学广播电视系_Silesia University's Radio and TV Department in Katowice/BAAS Arquitectura

回念：建筑嵌入、连续性和变化_Palimpsest: Architectural Insertions, Continuity and Change/Gihan Karunaratne

l Insertions,
nd Change

建筑师和设计师认为，与历史背景的接触不是一个限制，而是一个机会；当代建筑可以作为一个新的层次与之共存。它在建立未来遗产方面也发挥了重要的作用。在历史大环境之下进行建筑嵌入应该增加遗产建筑的价值，而不是贬低它们或破坏它们的特性和身份。

在历史大环境下，人们对什么样的发展方式是恰当的这个问题抱有不同的看法。新的建筑扩建结构及其具有的象征性是对它们那个时代的思想、技术、材料和建筑学词汇概念的视觉记忆。城市必须发展，因为城市是建筑和创新蓬勃发展的根本所在。它们应该考虑到建筑的创造性，并采用当代的结构和技术，这些结构和技术可以推动遗产的创造、促成连续性和最终变化的实现。[1,2]

Architects and designers perceive that engaging with a historical backdrop is not a constraint but an opportunity; a contemporary architecture can coexist as a new layer. It also performs a role in establishing the heritage of the future. Architectural insertion in a historic context should add value to heritage buildings rather than devaluing them or destroying their character and identity.

There are different opinions on what development is appropriate within a historic context. New architectural additions and typologies serve as the visual memory of ideas, technology, materials, and architectural lexicon of their period. Cities must be allowed to evolve, as cities are the fundamental places where architecture and innovation flourish. They should allow for architectural creativity and embrace contemporary structures and technologies which can add to the role of creating heritage, continuity and change.[1,2]

回念：建筑嵌入、连续性和变化

Palimpsest: Architectural Insertions, Continuity and Change

Gihan Karunaratne

纵观历史，建筑标志着时间的流逝，也见证了技术的变革。它反映了当时的社会政治、经济、艺术的条件和类型，以及知识和技术能力的构建状况。但是，在满足当代需求的同时，它还应该能够在规模、节奏和美学方面适应未来的使用的需求。

建筑物、城镇和城市都会随着时间的推移而变化，当然，历史上重要的建筑物和地区及其周边也会有所发展。如果我们确认一处历史遗迹是值得特别保护的，那么我们必须以特别小心的方式来进行。在这种情况下，建筑设计所扮演的角色应该是能够丰富现有的建筑遗产，而不是对其进行破坏，这一点至关重要。

从本质上讲，保护的目的是要保护一个具有历史意义的区域，同时要针对当前的要求进行改变，甚至尽可能做到对未来的情况进行预测。针对历史场所或文物遗址采取何种适当的设计方法，有着不同的理论：是应该采用历史风格，在材料和建筑技术上进行复制，还是应该像许多战后的现代主义开发项目一样，与传统的发展方式彻底决裂。还有一种折中的解决方案——新的嵌入结构，模仿历史特征，但使用当代材料和方法进行构建。

实际上，我们可以探索各种各样的发展途径，无论是传统的还是现代的方法，可能同样奏效。尽管如此，还是必须对建筑场地和周边大环境的特征和状况有深刻的认识，并知晓它们的独特之处，或者至少要知道这些历史建筑得以保存的价值所在。一般来说，设计必须是专门为这个特定的建筑地点环境而进行的。归类法或"照搬套用"的响应方法自然无法改善或者维持这个具有特殊意义的场地。

当我们在历史建筑中添加新的层次时，成功的设计将会迎接过去的挑战。历史大环境不是约束，而是机遇：过去和现在可以相互丰富。当创造力与理解力和敏感性相结合时，旧有的建筑结构会与新的扩建结构共同获得提升。[1]

Throughout history, architecture marks the passage of time and testifies to changes in technology. It reflects the socio-political, economic, and artistic conditions and genres of its period, and the state of building knowledge and technical capacity. But as well as meeting contemporary needs, it should be able to adapt, in terms of its scale, rhythm and aesthetics, to future use.

Buildings, towns and cities change over time, and development of course occurs in and around buildings and districts of historical importance. If a site of historical significance has been deemed to deserve special protection, then development must proceed in an especially careful manner. It is essential that the role architecture performs in this context is one that enriches, rather than robs, existing built heritage.

In essence, conservation seeks to preserve the character of a historic place while accommodating changes that respond to present demands and even anticipate – as far as is possible – future circumstances. There are contrasting theories about what is an appropriate design approach to historic or heritage sites: should a historic style be replicated in materials and building techniques, or should there be – as with many post-war Modernist developments – a complete break with traditional way of development. And then there is the halfway-house solution – new insertions that emulate historical character but are constructed using contemporary materials and methods.

In reality, a variety of development avenues may be explored, and one response – traditional or modern – may be as valid as another. Be that as it may, it is imperative that there is a profound recognition of the special character and qualities of site and context, and an understanding of what makes them unique, or least sufficiently significant to save. In general, a design response must be specific to the particular site context. Generic or "off the shelf" responses naturally fail to enhance or sustain sites of special interest.

When adding new layers to a historic fabric, successful designs meet the challenge that working with the past creates. Rather than acting as a restraint, a historic context presents opportunities: the past and present can be mutually enriching. When creativity is combined with understanding and sensitivity, old and new additions are enhanced.[1]

The following examples demonstrate an understanding and respect for the past combined with a high level of creativity and confidence in the present. They demonstrate that heritage sites can continue to have a meaningful role in the future.

以下的案例展示了在当前创造力和自信心无限高涨的情况下对过去的理解和尊重。它们表明，遗产地段在未来可以继续发挥有意义的作用。

该建筑场地位于香港的心脏地带。它拥有层次丰富的历史，但在目前的工程进行之前，公众不能进入参观。该场地由16座历史建筑组成，包括三座已获官宣的纪念性实体建筑——前中区警署、法院大楼和维多利亚监狱。这个开发项目是一个独特的"庭院"，位于世界上人口最密集的城市之一的中心地带。

在这个项目中，赫尔佐格&德梅隆将修复后的殖民建筑和高耸的立方体结构结合在一个多层复合建筑中。在研究实体结构时，为了能使它恢复到原始状态，他们非常谨慎小心。他们在设计中采用了一种大胆豪放而多元的方式，并在大馆中心的多元文化项目中得到了体现，该文化项目主要关注传统、当代和表演艺术，同时也以音乐、文学和电影为特色。该中心是一个用于文化交流和休闲的聚集地；它是一个文化锚定点，也是一个社会孵化器，将好几代人融合到了一种现代建筑类型之中。

这座建筑是用建筑钢桁架和特殊的混凝土饰面组合而成的。一段颇有特色的螺旋楼梯连接着所有的画廊。建筑外部覆盖着一层定制的立面系统，由可回收的铸铝制成，像乐高积木一样互相扣紧。这个几何形体取自现场周围一些现有的砖石结构，通过在整个立面周围创建砖一样的特征来建立文脉关系。这种建筑表现手法和材料的使用，使这座建筑在一组历史砖石建筑中显得与众不同。

该项目还考虑到了香港的亚热带气候。屋顶砖的几何形状被特别设计成一种遮阳装置，可以减少阳光的直射。立面单元的特殊孔隙度、图案的设计和表现方式由建筑内活动的功能和环境要求所决定。随着阳光穿透进来，下一个保护层上的拉伸织物会将漫射的环境光线洒进

The site is embedded in the heart of Hong Kong. It has a layered history, but before the present works were carried out it was not publicly accessible. The site comprises 16 historic buildings, including three declared monuments – the former central police station, Court House and Victoria Prison. The development is a unique "courtyard" in the center of the one of the densest cities in the world.

In this project – a multi-layered compound combining restored colonial buildings and a soaring cubic structure by Herzog & de Meuron – meticulous care has been taken in researching the physical fabric in order to return it to its original condition. A heroic yet pluralistic approach was taken in the design, and is reflected in the Tai Kwun Center's diverse cultural program, which focuses primarily on heritage and contemporary and performing arts, but also features music, literature and cinema. The center is a place to gather for cultural exchange and leisure; it is a cultural anchor and a social incubator which brings together and integrates multi-generations in a modern typology.

The building was constructed using a combination of architectural steel trusses and special concrete finishes. A featured spiral staircase connects all the galleries. The exterior is clad with a bespoke facade system made from recycled cast aluminum which interlocks like Lego blocks. The geometry was taken from some of the existing masonry work around the site to establish a contextual relationship by creating brick-like features all the way around the facade. This architectural expression and materiality set the building apart as new insertion among a collective of historical masonry blocks.

It also addresses Hong Kong's sub-tropical climate. The geometry of the rooftop bricks was especially designed to perform as a sun-shading device that cuts the direct sunlight. The specific porosity, patterning and expression of the facade units are informed by the functional and environmental requirements of activities pursued within the building. As the sunlight penetrates, stretched fabric on the next protective layer admits diffuse ambient light into the gallery spaces.[3]

This is an architecture which is well composed in relation to the different scales of the heritage buildings, and changes in volume and proportion. It works well within its historic context. The use of tactile materials, and the various tectonic qualities of these materials are consistent with the goals of making places expressive of construction and creating a vital architectural spatial narrative.

The Federico García Lorca Center (p.122) by MX_SI (Mendoza Partida + BAX studio) is located within the historical urban

画廊空间。[3]

这个建筑设计与遗产建筑的不同体量很好地相互呼应，并在体量和比例方面做出了改变。它在其特有的历史背景下运作良好。触觉材料的使用，以及这些材料的各种构造特征，都符合以下目标：使场所具有建筑表现力，并创造出一种非常重要的建筑空间叙事。

由MX_SI (Mendoza Partida+BAX Studio) 设计的费德里科·加西亚·洛尔卡中心（第122页）位于格拉纳达的历史城市环境中。这是一种微妙而精确的建筑干预手段，将文化设施嵌入到一个复杂而亲密的本土建筑类型之中。

该建筑有一个主要的立面，它结合了罗马广场的氛围，加以延续，并通过延伸入口和模糊内外边界的方式将其纳入室内项目。主要的设计方法是在公共和私人的许多方面进行干预，并创建一个与大环境持续有关的对话模式，这也是一种有着丰富叙事性的建筑。

在屋顶层，建筑保持较小的规模，因为它以碎片的方式融入到周围的城市结构之中。它源于雕塑的空洞，这也是连接城市环境和建筑形式的门槛。因此，主入口成为一个大入口，在这里，自然光和阴影之间的嬉戏被引入到了一个被空间关系彻底激活的室内空间。对自然光的对比处理正是这座建筑的精髓所在。

建筑叙事是一种分段讲述故事的方式，记录了城市空间的所有洞口，以及识别各种关系的亲密空间。这些序列的整合，以及通过使用几何结构实现的光影对比，创造了内部和外部空间之间的连续对话。在建筑的开放式平面图中，支撑构件仍然布置在周边。混凝土结构和建筑物的几何结构紧密相连；混凝土不仅用作支撑部件，也用作装饰材料。

建筑材料最重要的特点之一就是一定要轻。费德里科·加西亚·洛尔卡中心的材料色调受到了安东尼奥·罗塞蒂的雕塑作品《秘恋》的启发。它由布满纹理、等级标号不同的混凝土结构组成，表现了这种材料的透明度、亮度、粗糙度和重量。总的来说，这座建筑散发出一种温暖之感，并成功地与周围环境融为一体。[4]

context of Granada. It is a delicate and precise intervention that inserts a cultural facility inside a confined and intimate domestic typology.

The building has a primary facade which integrates and continues the Plaza de la Romanilla and brings it into the interior program by extending its threshold and blurring the boundary between inside and outside. The main strategy is to make interventions at many scales, public and private, and create a continuous and connected dialogue with the context – an architecture rich in narrative.

At roof level, the structure retains a smaller scale as it fragments to blend into the surrounding urban fabric. It derives from the sculpting void, the threshold that connects the urban context with the architectural form. Thus, the main ingress becomes a grand entrance where a play of natural light and shadow introduces an interior activated with spatial relationships. The contrasting treatment of natural light is the very essence of this building.

The architectural narrative is a story told through sections, recording all the openings of the urban spaces as well as intimate spaces which identify relationships. The integration of the sequences, and contrasts between light and shadow achieved through the use of geometrical structures, create a continuous dialogue between interior and exterior. In the building's open plan, the supporting elements remain around the perimeter. The concrete structure and the building's geometry go hand in hand; concrete is used not only as a supporting element but also as the finishing material.

One of the most important building materials is light. The material palette of the Federico García Lorca Center is inspired by the Antonio Rossetti sculpture Secret Love. It consists of textured and variously graded concrete construction which expresses and represents transparency and lightness, as well as roughness and heft. Altogether, the building exudes warmth and connects successfully with its environment.[4]

The Silesia University's Radio and TV Department (p.138) designed by BAAS Arquitectura with the local partners, Grupa 5 Architekci and Maleccy Biuro Projektowe is embedded in a built-up area of the city of the Polish city of Katowice. The site included a vacant building which the client initially intended to demolish.

Instead, the realized scheme preserves the existing building and includes an extension to it, while conserving the

1. Contemporary Architecture In Historic Urban Environments By Susan Macdonald
2. Perfection or Pastiche? New Buildings in Old Places - Professor Simon Thurley
3. https://video.arup.com/?v=1_xd7xj815
4. MEXTRÓPOLI 2016, MX_SI - Mara Partida y Héctor Mendoza, Arquine, Published April 1, 2016.

由BAAS建筑事务所与其当地合作伙伴Grupa 5建筑事务所和Malecy Biuro Projektowe事务所共同设计的西里西亚大学广播电视系（第138页），被嵌入到波兰卡托维兹市的城区。这个建筑地点包括一栋闲置的建筑，而客户最初本打算要拆除它的。

现在的方案恰恰相反，方案中保留了现有建筑，并对其进行扩建，同时保留了现有结构的特征。该项目还包含了一座占据了内部街区、高度较低的建筑。这种干预强调了中央庭院的突出性，并使这里的城市结构具有连续性。该设计对现存建筑的美观性很敏感，并通过在其上方安装一个由格子砖结构制成的抽象体量来充分利用其物质性和视觉质量，而格子砖结构也跟周边的建筑相契合。新立面由与现有立面相同的砖砌结构组成，因此提供了与现存建筑的视觉联系。开放的砖砌结构给空间带来了一种抽象而又独特的特征。在上部楼层，墙壁会过滤光线，在视觉上孤立的内部空间中营造了一种安静、专注的氛围。该建筑设计对地理、地貌、文化等做出了回应，找到了适合自己的美学表达方式。

大楼的主入口似乎是从主体建筑中凿出来的。玻璃入口大堂将街道与已改建为图书馆的现存建筑以及新结构的内部连接起来。主入口使得中庭周围的交通流线和楼层之间的连接成为可能。建筑的主入口和内部庭院之间有一个清晰的连接，这有助于将外部街道引入建筑，反之，也将大学延伸到了外部环境中去。

新大楼占据了整个地块，同时中间挖空，形成了一个中央庭院，所有在新建院系大楼的演播室和演讲室周围开展的社交活动都将在这里举行。上述谈到的这三个建筑案例都展示了一种处理现存建筑结构的方法，该方法试图让历史建筑占据前景，而新扩建的部分将进入背景。这种方法强调了原始结构，并与周围的大环境创建了一种崭新的关系。对于许多新建筑过于注重自我展示的设计，这些案例无疑都是诱人的解毒剂。无论是在特定的建筑规模上还是在城市中，它们都表现出一种更加安静、微妙，更加专注也更尊重过去的工作方式。

character of the existing fabric. The project also accommodates a lower-height building occupying the interior block area. This intervention emphasizes the prominence of the central courtyard, and gives continuity to the urban fabric.
The design is sensitive to the existing building's aesthetics and takes advantage of its materiality and visual qualities by erecting on top of it an abstract volume made out of a brick latticework, which follows the neighbor's section. The new facade is composed of the same brickwork as the existing facade, so providing the visual connection with the existing building. The open brickwork gives an abstract and unique character to the space. On the upper level, the wall filters the light, imparting an atmosphere of silence and concentration in the visually isolated inner space. This is an architecture which responds to geography, terraform and also to culture to find its aesthetic lexicon.

The main entrance to the building seems chiseled out from the main volume. The glazed entrance lobby connects the street with the existing building, which has been converted into the library, and with the inner part of the new structure. The main entrance allows the possibility of circulation around the atrium and connection between the floors. There is a clear connection between the building's main entrance and the inner patio which helps to bring the external street into the building, and the university to the outside.

The new building occupies the whole plot and at the same time hollows out a central courtyard and becomes the key element for all the social activities taking place around the studios and lecture rooms at the new university department. All three of these architectural examples demonstrate an approach to working with existing building fabric that seeks to let the historic building take the foreground, while the new addition steps into the background. Such an approach enhances the original and creates a new context around it. The exemplars are captivating antidotes to the "look at me" designs of many buildings. They demonstrate a much quieter, subtle, and more focused and respectful approach to working with the past, both on the scale of a specific building and of the city.

大馆传统与文化中心
Tai Kwun, Center for Heritage & Arts
Herzog & de Meuron

前中区警署旧址所在的区域，包括中央裁判司署和维多利亚监狱，是位于香港岛商业中心的一片有围墙的文物建筑。该建筑场地于2006年停止使用并被腾空，留下了一片空地和一组独特的建筑。它们曾经作为法律和秩序的突出象征矗立在山坡之上，从这里可以俯瞰港口，现如今却成了在商业和住宅高楼林立的森林中保持开放和平静的一片城市绿洲。在这里，人们可以聚集在一起进行文化交流、休闲活动和休息。

该建筑由两个大型庭院组成：阅兵场和监狱庭院。阅兵场周围是该遗址最具历史意义的几座建筑，形成了一个正式的开放区域，为公共娱乐、活动、餐馆以及零售场所提供了宽敞的空间，同时，还可以成为较小规模的文化和教育空间。

监狱庭院由一片建筑风格粗犷的禁区转变为一个开放的文化公共空间，专门为文化项目提供了场所。两个新建的悬臂结构“漂浮”在周边的花岗岩墙之上，设计独特，并仔细嵌入了现有建筑结构的元素之中。

通过在墙壁上方进行悬臂式的设计，并与周围历史建筑的留出一个有保护性的空地，建筑面积实现了最大化，同时又保持在分区结构的范围之内。通过增加体量，新的公共空间和交通流线空间被设置在

建筑的下部楼层——为社会活动以及从阿巴思诺特路到老贝利街的新东西向人行通道提供了宽敞的受保护空间。同时，这些建筑在建筑复合体的两端成为新的标志，将人们的注意力吸引到这个城市原来被封闭的部分。

位于场地西南部的老贝利翼楼的设计与相邻的F厅的适应性再利用设计紧密结合。它们共同容纳了当代艺术功能。老贝利翼楼的庭院入口设计是由悬停在历史护墙上方的建筑体量所定义的。项目设计使该入口区域免受雨淋日晒，并作为F厅和老贝利翼楼的交通流线枢纽和门厅。

F厅是艺术展览或其他特殊活动的多功能画廊空间。许多原来的建筑和建筑过去使用的痕迹都被保存了下来。相比之下，邻近的老贝利翼楼因为采用了无柱大展览空间的设计而具有最大限度的灵活性。屋顶的天窗为顶部的画廊空间提供了自然采光。置身三层餐厅的露台，人们可以俯瞰嵌入城市景观中的整座建筑纹理丰富的屋顶景观，感受它的富丽堂皇。

新阿巴思诺特翼楼作为一座全新的地标性建筑，位于场地的东南角。这个体量悬停在阿巴思诺特路护岸墙的上方，打造了一个带顶的公众室外集会、电影院和表演空间，还有一个大楼梯，台阶可用于非正

JC CONTEMPORARY

式的座位。在阿巴思诺特翼楼的两侧是古老的监狱建筑、D厅和E厅。它们共同形成了这个独特的户外场地的内部立面。高度限制也得到了充分利用，使得上方产生了两个更为宽敞的功能空间：一个8m高的多功能空间，带有可进入的设备网格天花板，以及一个机械设备室，里面装有冷却塔和整个场地的设备。

两栋新建筑均采用铸铝覆盖，从规模和比例上参考了整个场地周围具有特色的边界护墙上现有的花岗岩块元素。百分之百回收铝材的使用提供了独特的建筑表现形式和材料特性，将新扩建的结构作为历史砖石建筑中的嵌入部分分开，同时解决了诸如结构支撑、遮阳以及在香港的亚热带气候下的防雨等问题。立面单元的特殊多孔性、图案和表现方式均由室内的功能和环境要求决定。铸铝单元的粗糙度和纹理打破了立面表面的一致性，这有助于减少白天的反光和眩光。晚上，建筑发出的光有一部分被立面遮挡，用这种方式表达了室内活动所蕴含的生命力。

The former Central Police Station site, including the Central Magistracy and the Victoria Prison, is a walled compound of heritage buildings at the commercial center of Hong Kong Island. The site was decommissioned and vacated in 2006, leaving a set of open grounds and a collection of unique buildings. What once stood on the hillside as a prominent symbol of law and order, with commanding views to the harbor, has now inversely become distinct urban oasis of openness and calm within a forest of commercial and residential high-rises; a place of gathering, cultural exchange, leisure, and respite.

The compound is defined and structured by two large courtyards: The Parade Ground and the Prison Yard. The Parade Ground is surrounded by several of the site's most historic buildings, resulting in a formal open space with generous room for public recreation, events, restaurants and retail attractions, as well as smaller-scale cultural and educational spaces.

The Prison Yard is transformed from a rough and forbidding area into an open cultural public space dedicated to cultural programming. Two new cantilevered volumes that "float" above the surrounding granite walls are conceived as distinctive but carefully inserted elements within the fabric of existing buildings.

By cantilevering above the walls and keeping a protective offset from the surrounding historical buildings, the buildable floor area is maximized while staying within the zoning envelope. By raising the volumes, new public and circulation spaces are created below – generous protected space for social activity and a new east-west pedestrian connection from Arbuthnot Road to Old Bailey Street. At the same time, these buildings become new markers at the two ends of the compound, bringing attention to a formerly closed-off part of the city.

The design of Old Bailey Wing, located to the south-west of the site, is developed closely alongside the adaptive reuse

西立面 west elevation

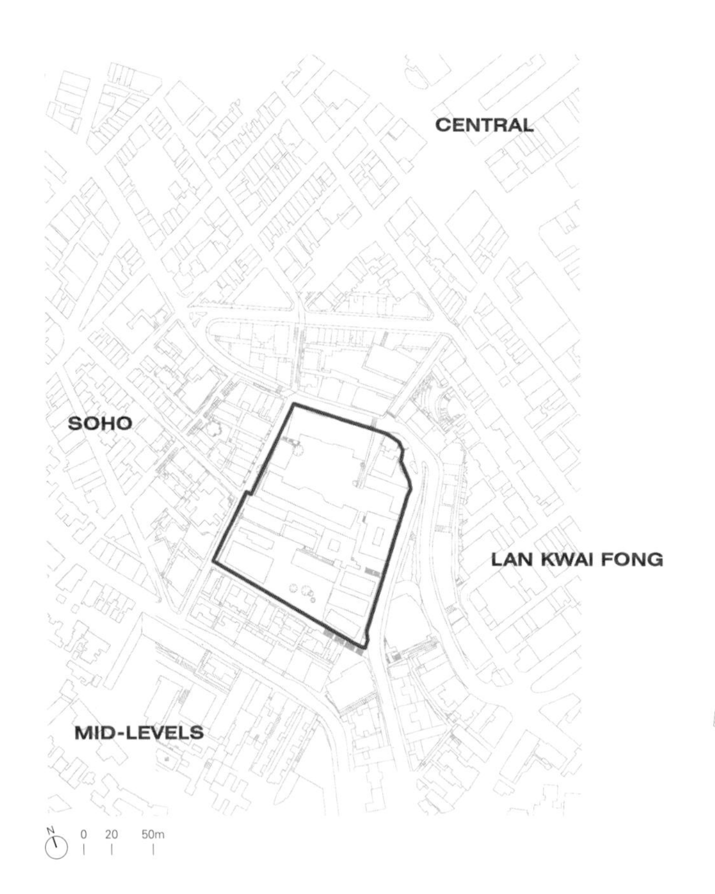

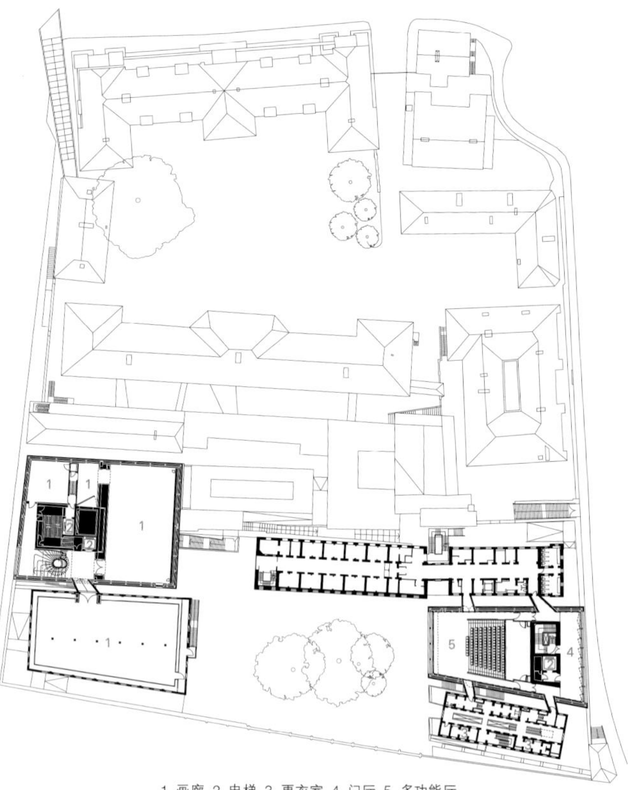

1. 画廊 2. 电梯 3. 更衣室 4. 门厅 5. 多功能厅
1. gallery 2. lift 3. dressing room 4. foyer 5. multipurpose hall
二层 first floor

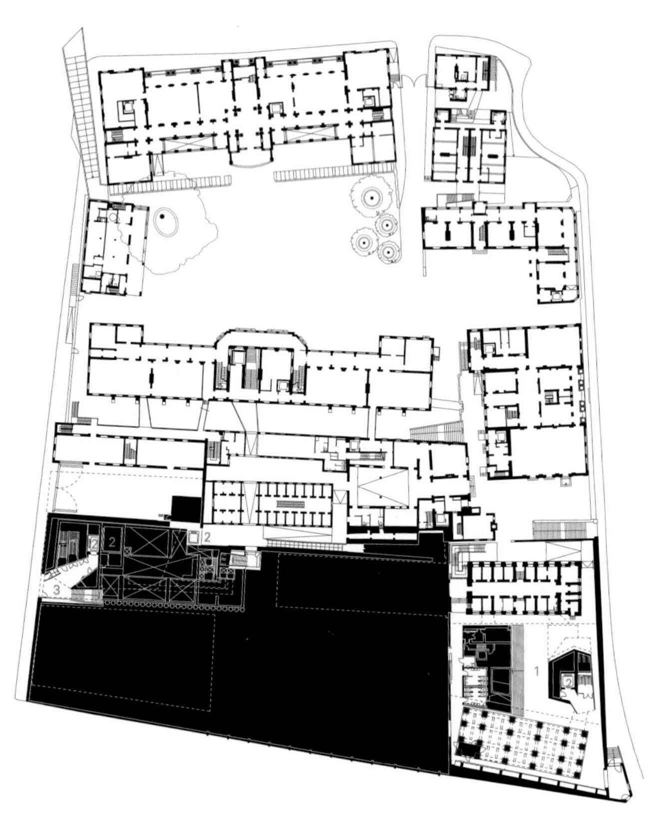

1. 洗衣场 2. 电梯 3. 前厅
1. laundry yard 2. lift 3. lobby
地下一层 first floor below ground

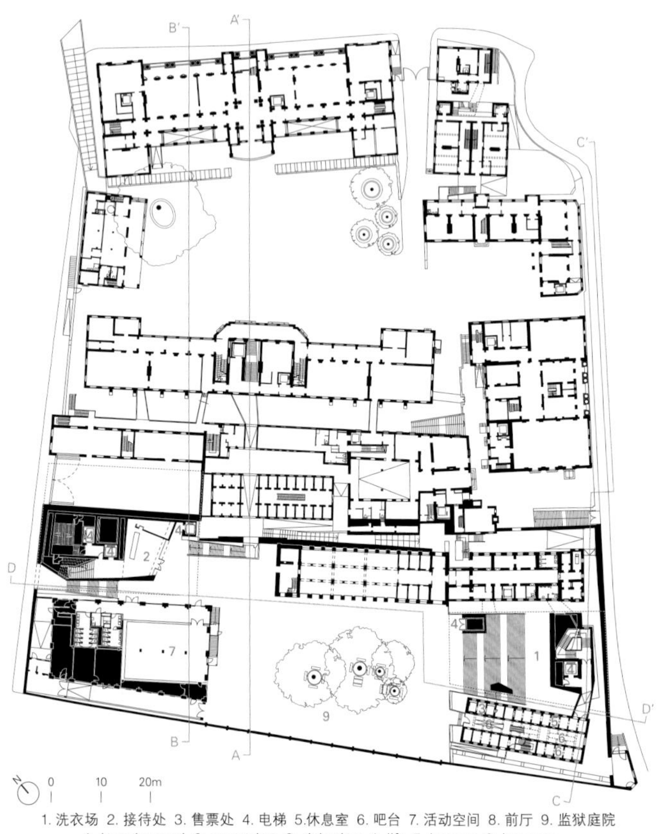

1. 洗衣场 2. 接待处 3. 售票处 4. 电梯 5.休息室 6. 吧台 7. 活动空间 8. 前厅 9. 监狱庭院
1. laundry yard 2. reception 3. ticketing 4. lift 5. lounge 6. bar area
7. event space 8. lobby 9. Prison Yard
一层 ground floor

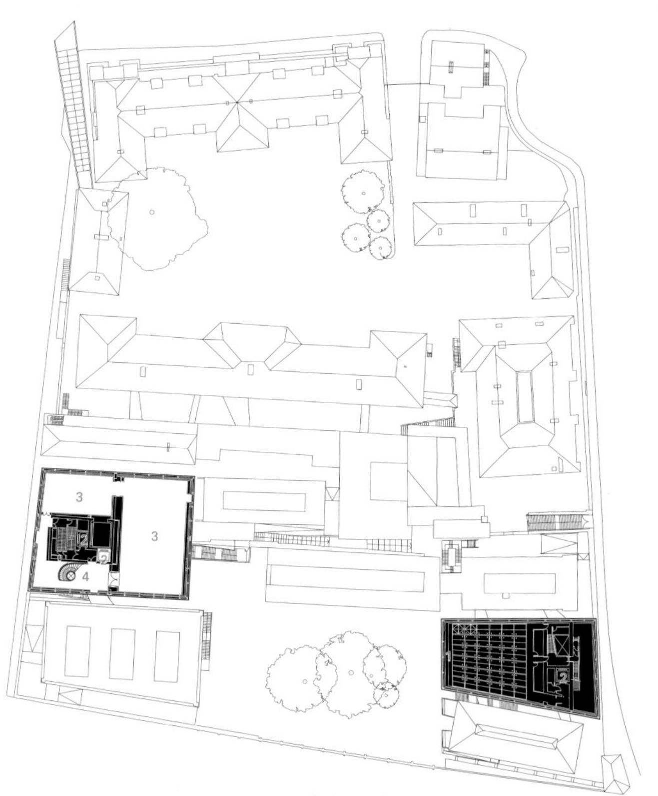

1. 设备天花板 2. 电梯 3. 画廊 4. 门厅
1. technical ceiling 2. lift 3. gallery 4. lobby
四层 third floor

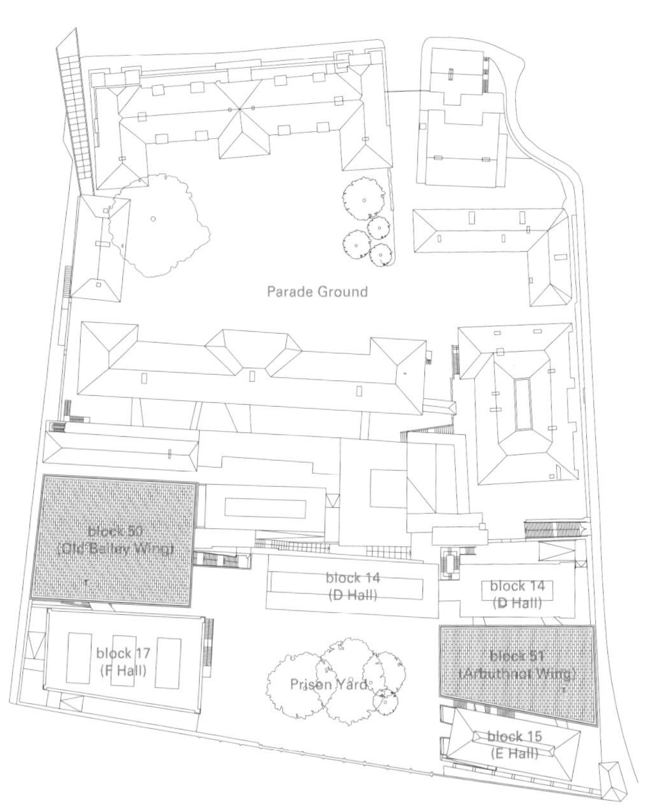

屋顶 roof

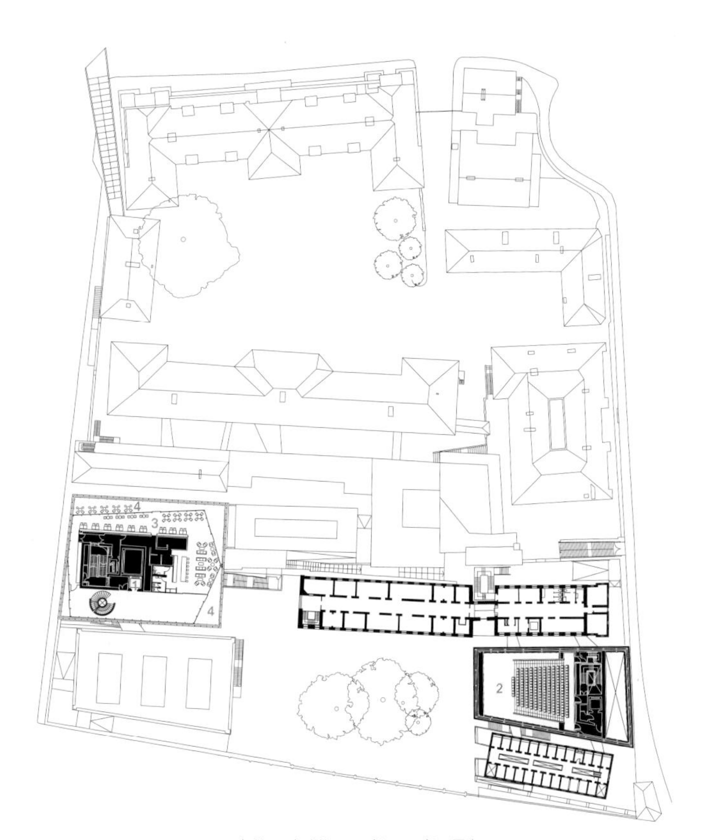

1. 电梯 2. 多功能厅 3. 餐厅 4. 餐厅露台
1. lift 2. multipurpose hall 3. restaurant 4. restaurant terrace
三层 second floor

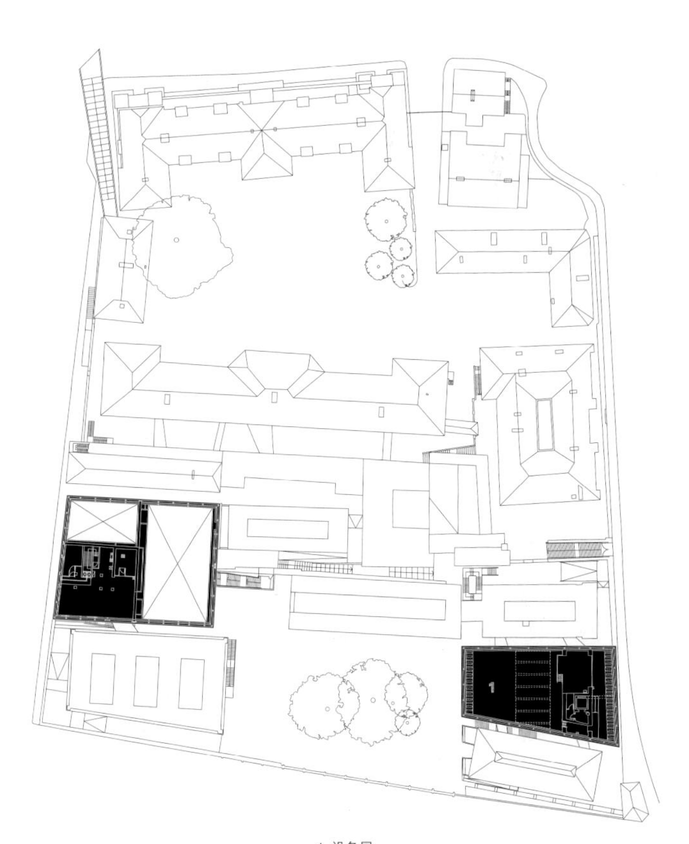

1. 设备层
1. technical floor
五层 fourth floor

座 E HALL 伍

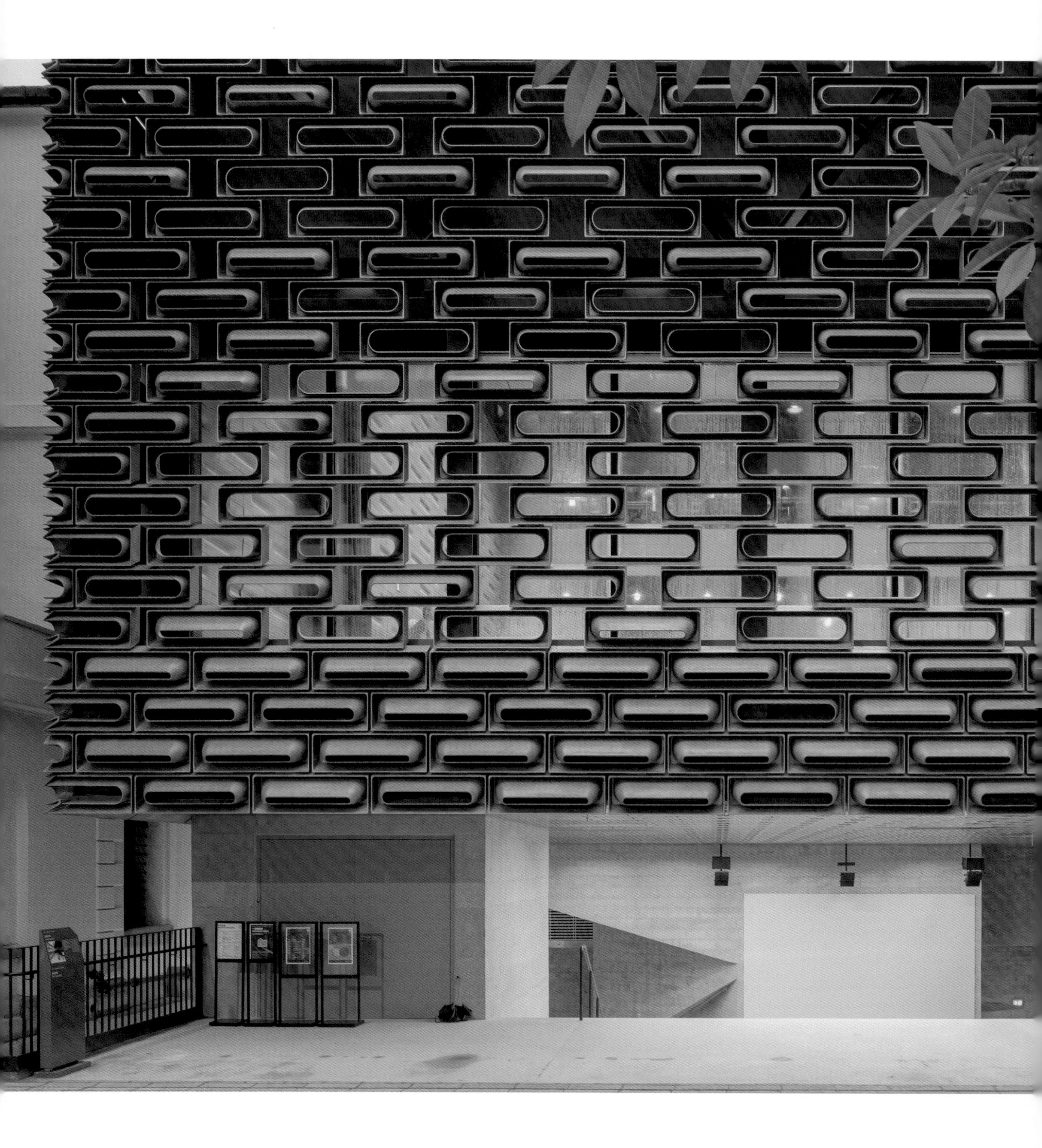

项目名称：Tai Kwun, Center for Heritage & Arts / 地点：10 Hollywood Road, Hong Kong, China / 建筑师：Herzog & de Meuron–Jacques Herzog, Pierre de Meuron, Ascan Mergenthaler (Partner in charge) / 项目团队：Associate, Project director–Vladimir Pajkic, Edman Choy; Project manager–Chi-Yan Chan; Associate–Raymond Jr. Gaëtan / 执行建筑师：Rocco Design Architects Ltd. / 保育建筑师：Purcell / 结构、土木工程、立面、防火、岩土工程+照明、安全、IT顾问：Arup / 声学工程、视听设计顾问：机电工程：Shen Milsom & Wilke Ltd. / 机电工程：J Roger Preston Ltd. / 施工技术员：Rider Levett Bucknall Ltd. / 交通规划：MVA
顾问：Art advisor–David Elliot; BEAM + Sustainability–Hyder Consulting Ltd.; Building physics–Transsolar Energietechnik GmbH; Environmental & Archaeology–ERM; Landscape–AECOM; Metallurgist–C M Whittington & Associates Pty Ltd.; Planning–Townland Consultants Ltd.; Retail Operation–Knight Frank; Scenography–dUCKS scéno; Signage–Marc & Chantal Design; Theatre advisor–Philip Soden (Hong Kong Academy of Performing arts); Tree Survey–Prof. C Y Jim (Hong Kong University) / 客户：The Hong Kong Jockey Club / 摄影师：©Iwan Baan (courtesy of the architect)

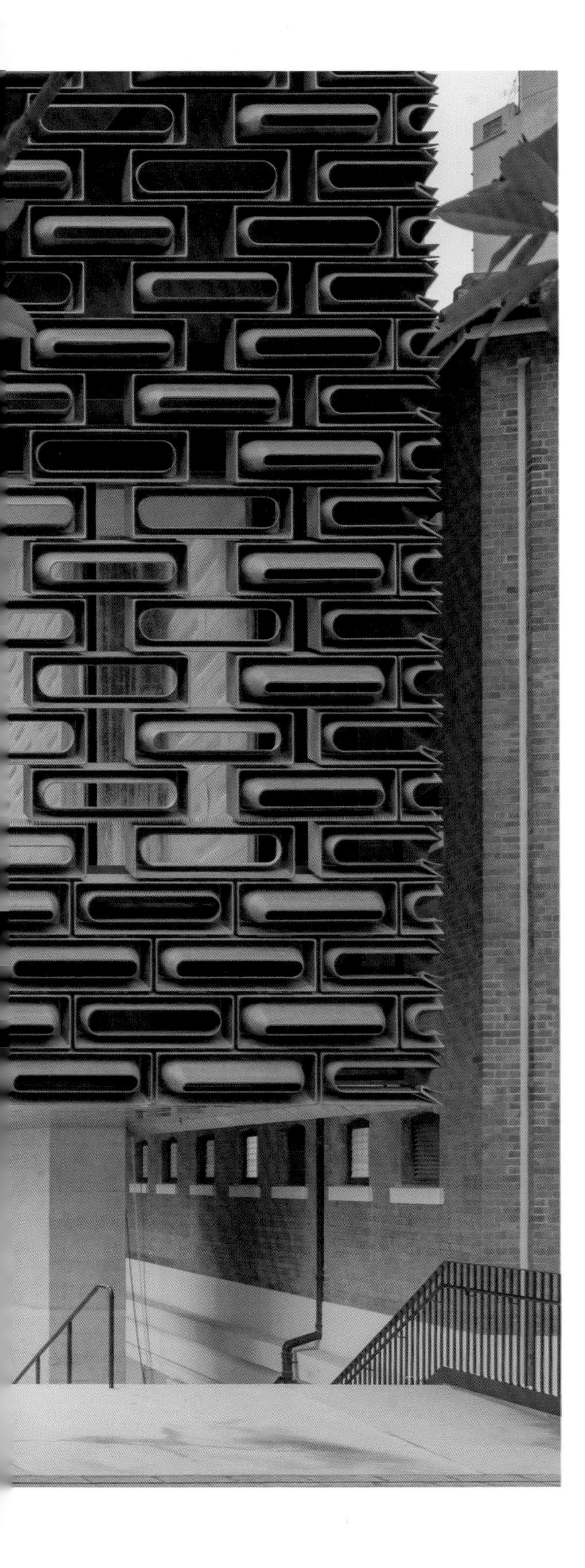

用途：old Bailey wing + F Hall－galleries;
Arbuthnot wing－multi-purpose hall;
Tai Kwun site－arts & cultural program, retail, food & beverage, public use & plant / 用地面积：14,500m² of which 4,000m² for open courtyard / 建筑面积：10,500m² of which 800m² for old Bailey wing and 550m² for Arbuthnot wing
总建筑面积：27,000m² / 建筑规模：twenty-four stories
设计时间：2006—2011 / 施工时间：2011—2018

of adjacent F Hall. Together, they function to house contemporary art. The courtyard entrance of the Old Bailey Wing is defined by the building's volume hovering above the historic revetment wall. This entrance area is protected from rain and sun, and acts as the circulation hub and foyer to both the F Hall and Old Bailey Wing.

The F Hall serves as a versatile gallery space for art exhibition or other special events. Much of the original architecture with its traces from the buildings' past use are kept and preserved. In contrast, the adjacent Old Bailey Wing feature large column-free exhibition spaces with maximum flexibility. Skylights in the roof will provide natural daylight into the top gallery space. The terrace of the restaurant on the second floor overlooks the rich and textured roofscape of the entire compound, embedded in the cityscape.

Sitting as a new marker on the southeast corner of the site is the new Arbuthnot Wing. The volume similarly hovers above the revetment wall along Arbuthnot Road, creating a covered public outdoor gathering, cinema and performance space with a large stair whose steps may be used for informal seating. On either side of the Arbuthnot Wing are old prison buildings, D Hall and E Hall, forming interior elevations for this unique outdoor venue. The height restriction is also fully taken advantage of, resulting in two more generous program spaces above: an 8m height multipurpose space with an accessible technical grid ceiling, and a mechanical plant room that houses cooling towers and equipment serving the entire site.

Both new buildings are clad with a cast aluminum, referencing the existing granite block elements of the characteristic bordering revetment wall surrounding the entire site in terms of scale and proportion. The use of 100% recycled aluminum provides a distinctive architectural expression and materiality, setting the new additions apart as insertions amongst the historical masonry buildings, and at the same time addressing issues such as structural support, sun shading, and rain protection in Hong Kong's subtropical climate. The specific porosity, patterning, and expression of the facade unit are informed by the functional and environmental requirements of the uses within. The roughness and texture of the cast aluminum units break down the facade surface, which helps to reduce reflectivity and glare during the daytime. At night, light emitted from the building is partially screened by the facade units, expressing the life of activities within.

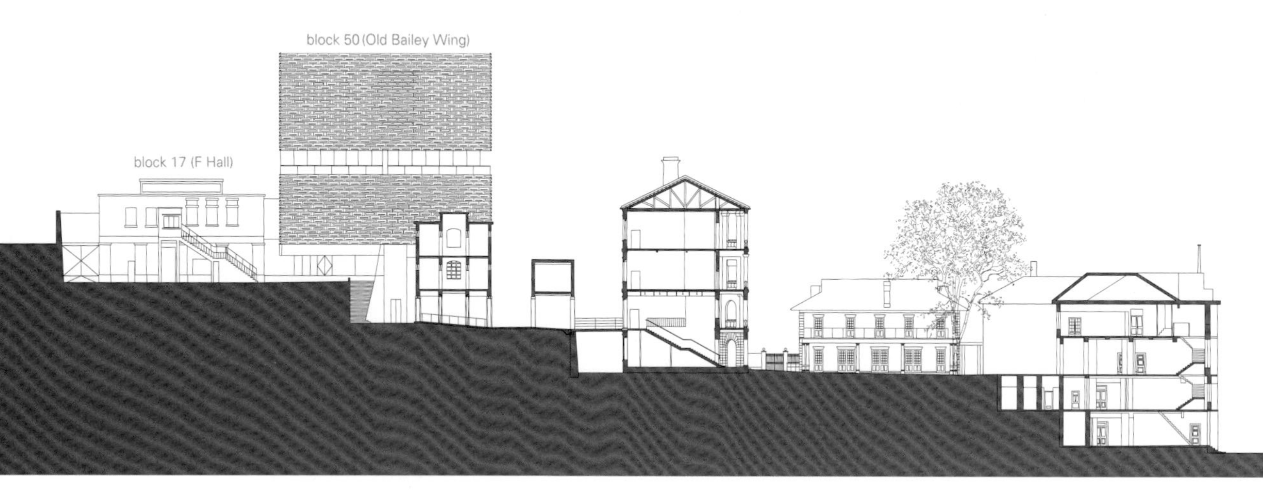

A–A'剖面图 section A-A'

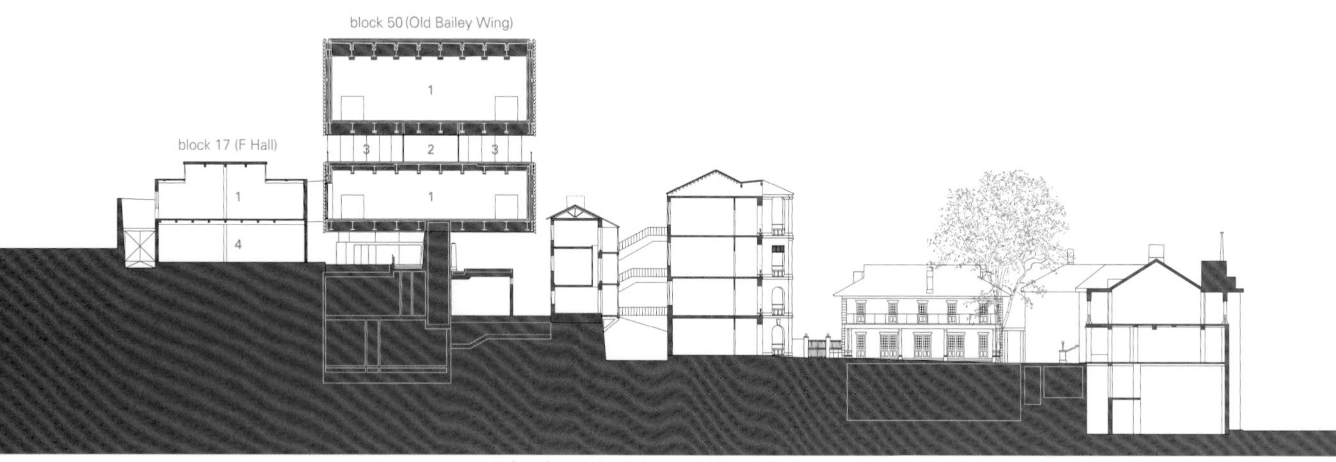

1. 画廊 2. 餐厅 3. 餐厅露台 4. 活动空间
1. gallery 2. restaurant 3. restaurant terrace 4. event space
B–B'剖面图 section B-B'

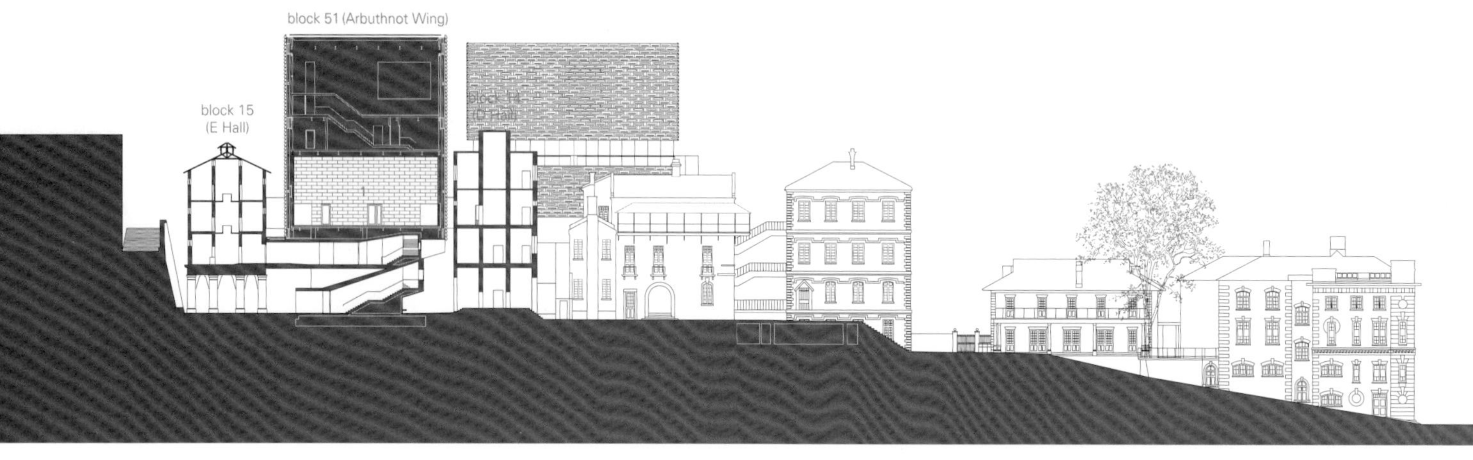

1. 门厅 1. foyer
C–C'剖面图 section C-C'

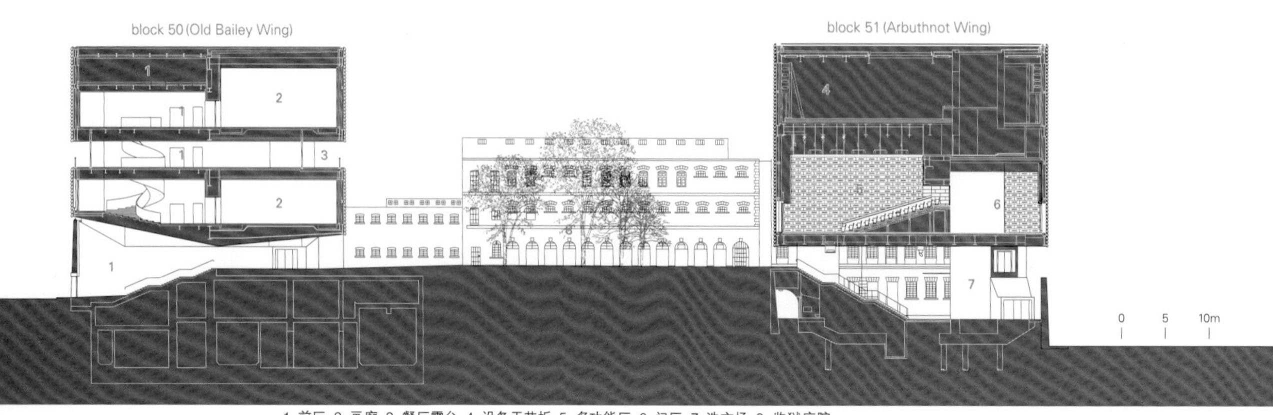

1. 前厅 2. 画廊 3. 餐厅露台 4. 设备天花板 5. 多功能厅 6. 门厅 7. 洗衣场 8. 监狱庭院
1. lobby 2. gallery 3. restaurant terrace 4. technical ceiling 5. multipurpose hall 6. foyer 7. laundry yard 8. Prison Yard
D–D'剖面图 section D-D'

格拉纳达费德里科·加西亚·洛尔卡中心

Federico García Lorca Center in Granada

MX_SI (Mendoza Partida + BAX studio)

Zorongo
Zorongo

位于格拉纳达的罗马广场上的费德里科·加西亚·洛尔卡中心，是为了纪念西班牙诗人、剧作家和戏剧导演费德里科·加西亚（1898—1936年）而建造的。他在西班牙内战开始之初就被民族主义武装力量处决了。

该中心致力于推广和传播这位诗人的作品，以及其他具有共同价值观的西班牙乃至国际当代艺术、文学、戏剧、音乐、研究和教育项目，特别是标志着诗人遗志所蕴含的捍卫自由的精神。

2005年举办的国际建筑概念设计大赛，旨在设计一座能容纳艺术、学术和社会活动的建筑。

大赛评委会——包括拉斐尔·莫尼奥、弗朗西斯科·达尔科、胡安·乔斯·拉韦尔塔、阿尔贝托·坎波·巴埃萨、胡安·卡拉特拉瓦、恩格尔·路易斯·吉恩、卡洛斯·阿尔韦迪、恩格尔·穆奥兹·卡德纳斯、路易斯·赫拉尔多·加西亚·罗约、因马库拉达·洛佩斯·卡拉奥罗和劳拉·加西亚·洛尔卡——最终选择了MX_SI建筑设计工作室。这是一家位于巴塞罗那的建筑公司，之所以胜出就是因为他们对巴塞罗那的遗产保护和地方身份有着很强的敏感度。

该设施成为一个崭新的文化地标，出现在公众面前，自然而又流畅地融入格拉纳达市的历史结构之中，具有显著的城市代表性。

新空间入口的设计大气磅礴——这是一个象征着费德里科·加西亚·洛尔卡中心向城市开放的入口，反之亦然。其建筑设计将设计标书中提到的门厅和公共设施融入到室外空间之中，成为整体环境的一部分。

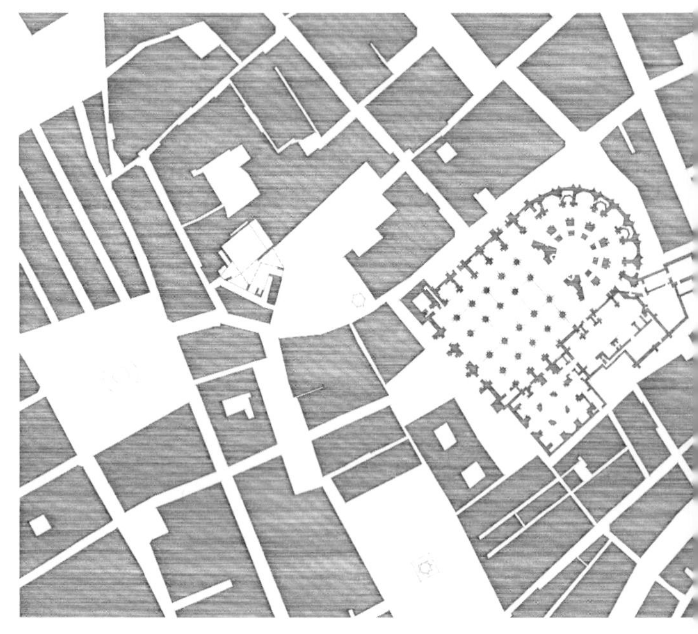

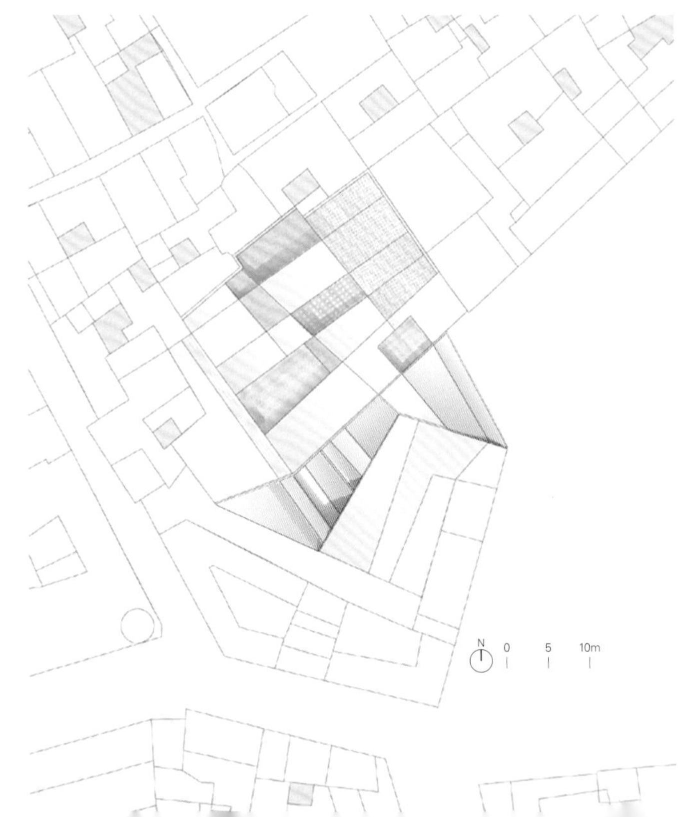

建筑构成了公共空间的一个自然部分，融入到整个城市环境之中，创造出一系列城市场景，使其所在的广场能够一直延伸到建筑当中去。

历史城市环境和建筑空间之间的界限通过延长入口而变得模糊，也就是我们上面说的朝向罗马广场的公共区域的入口。

设计策略包括将架空空间以不同的比例雕刻成一个宏伟的入口。通过这种方式，游客可以通过横梁和连桥覆盖的通道进入内部，横梁和连桥形成了互相照射的光影，从而创造出全新的三维空间关系。

反过来，干预设计自然而流畅地将广场与中心的地面结合在一起，这是一个没有任何结构元素的空间，可举办灵活多样的活动。

这个建筑群由两个建筑体量组成，建筑面积达4700m²，由一系列灵活的展示空间组成，这些空间可以承办各种文化活动。有400个座位的剧院配有隔声外壳和移动墙，从门厅区域就开始设置隔声装置，为音乐会的举办和戏剧的演出保证了必要的音质效果。

由两个高度不同的空间组成的图书馆，主要用于藏书和阅览室使用。在这里，人们可以看到一个悬挂的盒子结构，这是一个用铁皮包裹的档案馆，用于保护诗人的原始手稿。

地下室是一个宽敞的空间，包括自助餐厅、商店、办公室、储藏空间和设备区域。

新建的中心是用现浇混凝土建造的，这样处理使它的表面在视觉和触觉方面都会产生光滑的效果。在建造中使用了与天然石材相似的纹理，使得建筑与格拉纳达历史中心的建筑相呼应，使该中心成为社区的文化地标。

The Federico García Lorca Center, in Granada's Plaza de la Romanilla, was built in honor of Federico García Lorca (1898—1936), a Spanish poet, playwright, and theater director. He was executed by Nationalist forces at the beginning of the Spanish Civil War.

The center promotes and disseminates the poet's work, alongside other national and international contemporary art, literature, theater, music, research and education which share the values – in particular the defense of freedom – that mark his legacy.

The international ideas competition was launched in 2005 to design a building capable of housing the artistic, academic and social program of the center.

The jury – comprising Rafael Moneo, Francesco Dal Co, Juan José La Huerta, Alberto Campo Baeza, Juan Calatrava, Ángel Luis Gijón, Carlos Alberdi, Ángel Muñoz Cadenas, Luis Gerardo García Royo, Inmaculada López Calahorro and Laura García-Lorca – chose the project of MX_SI architectural stu-

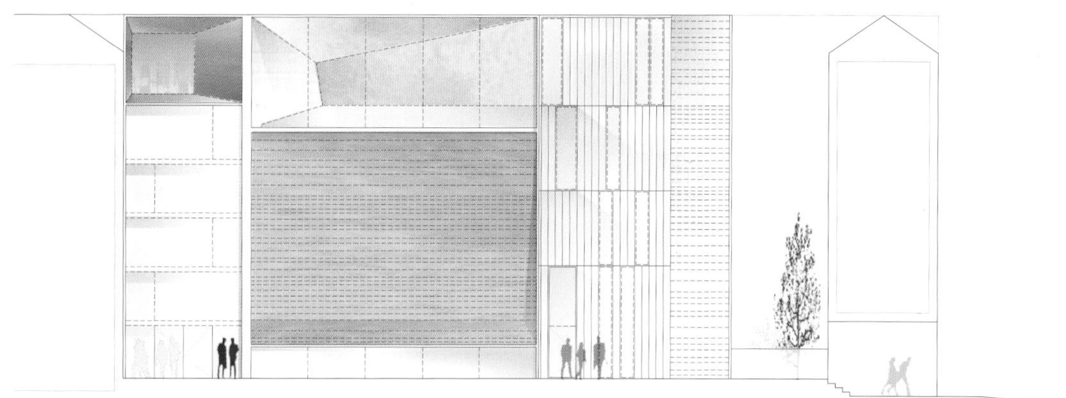

西南立面 south-west elevation

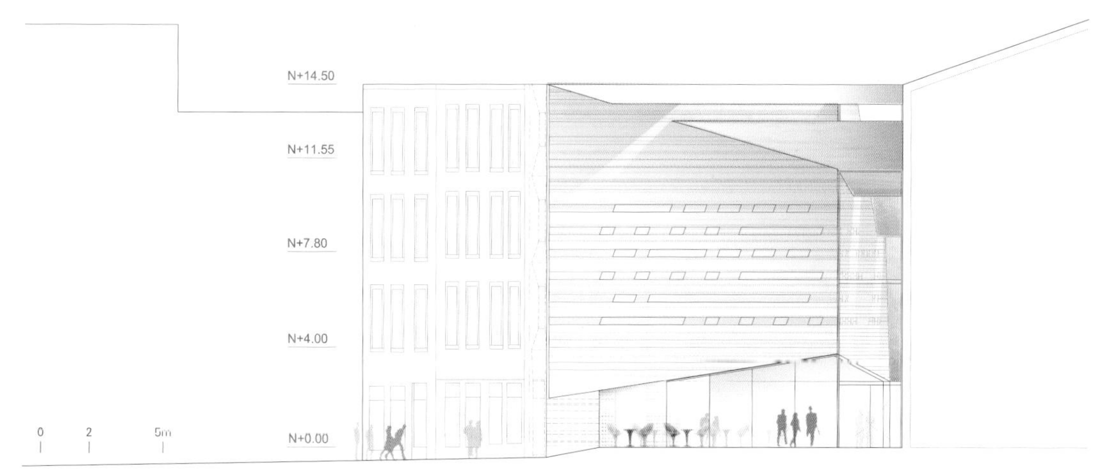

东北立面 north-east elevation

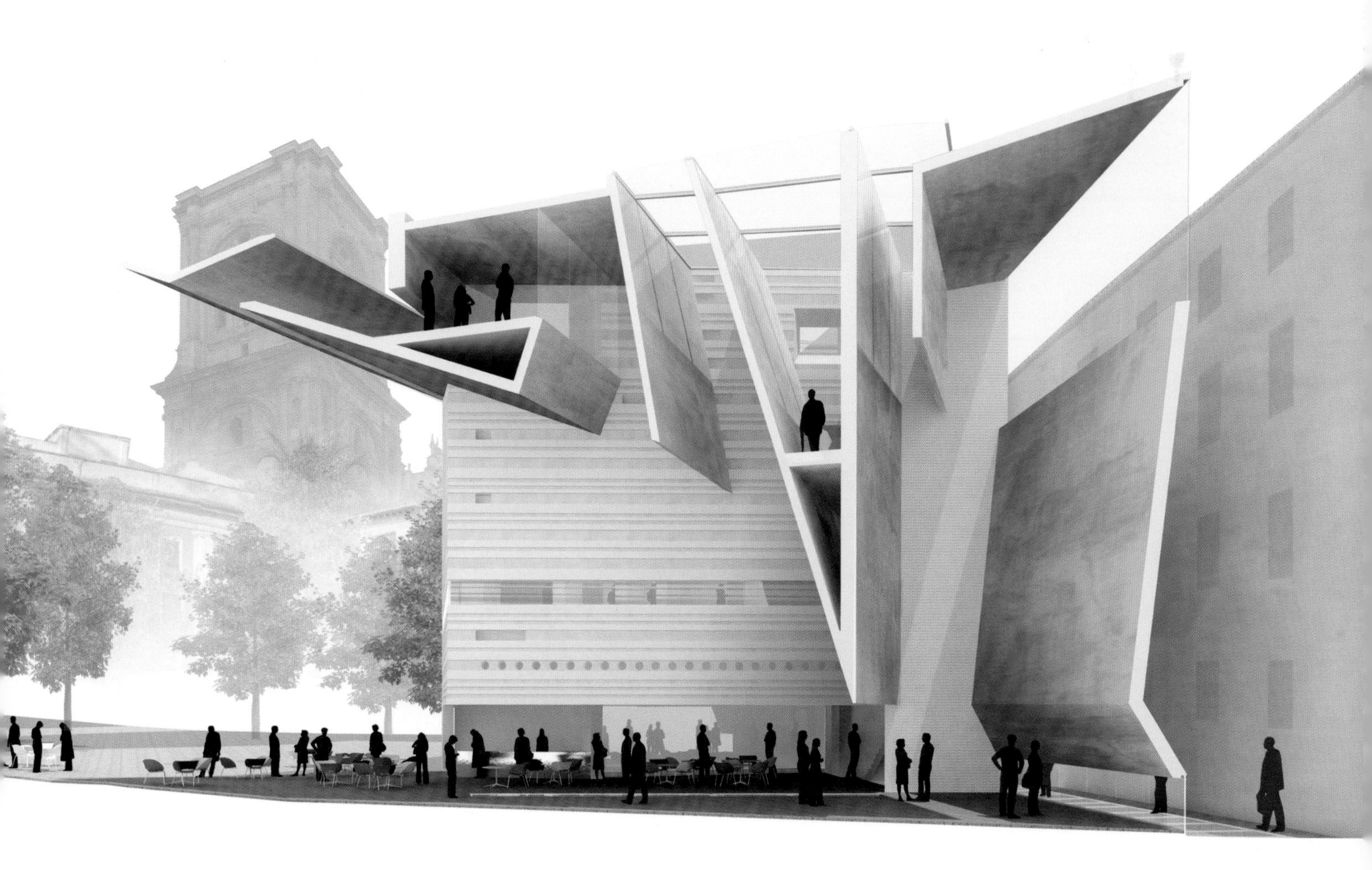

García Lorca
Centro Federico García Lorca
Bienvenidos / Welcome

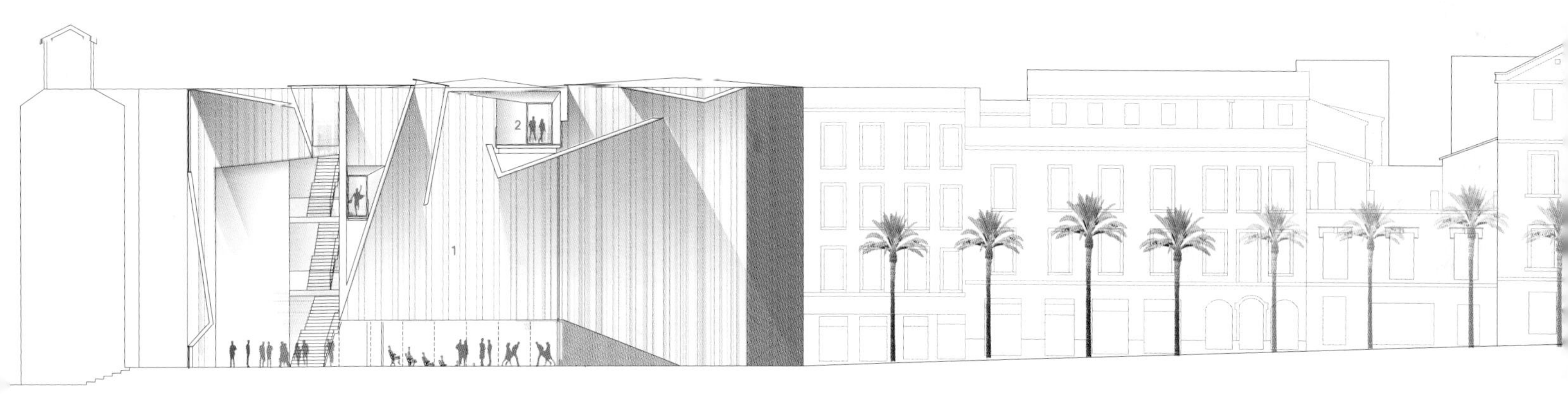

1. 门厅通道 2. 大厅 1. foyer-passage 2. hall
A–A'剖面图 section A-A'

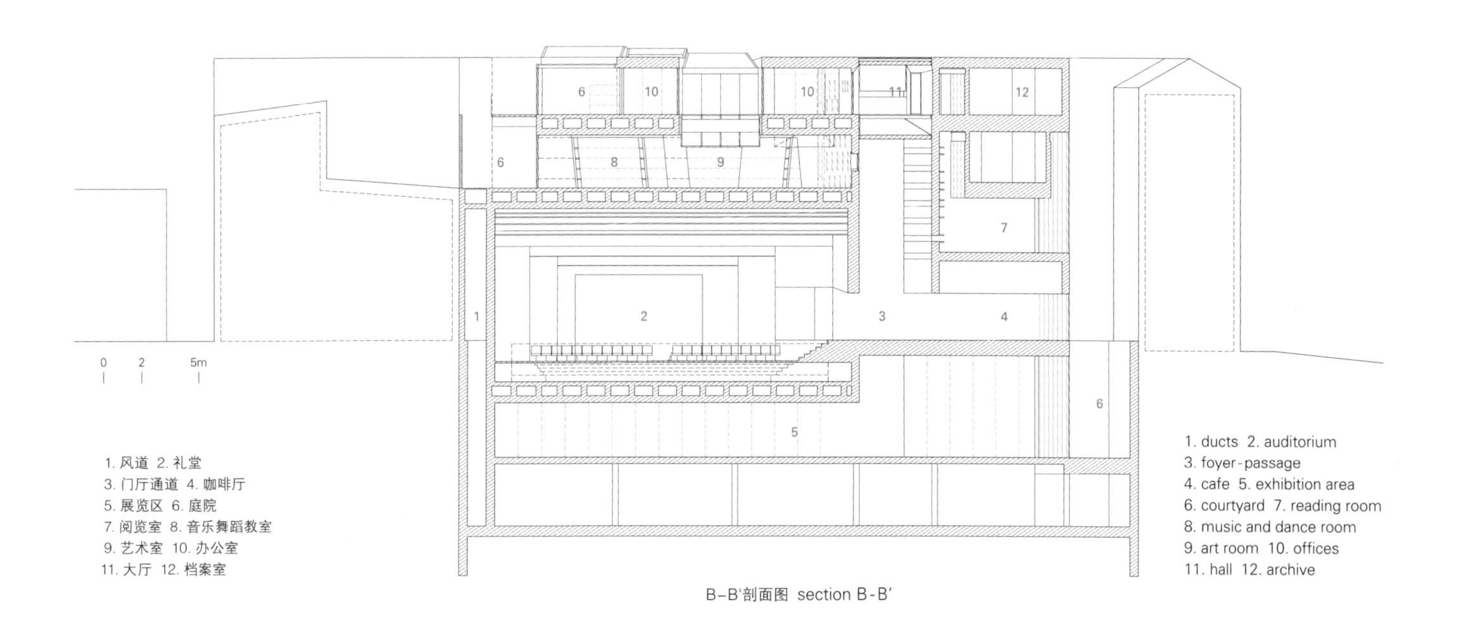

1. 风道 2. 礼堂
3. 门厅通道 4. 咖啡厅
5. 展览区 6. 庭院
7. 阅览室 8. 音乐舞蹈教室
9. 艺术室 10. 办公室
11. 大厅 12. 档案室

1. ducts 2. auditorium
3. foyer-passage
4. cafe 5. exhibition area
6. courtyard 7. reading room
8. music and dance room
9. art room 10. offices
11. hall 12. archive

B–B'剖面图 section B-B'

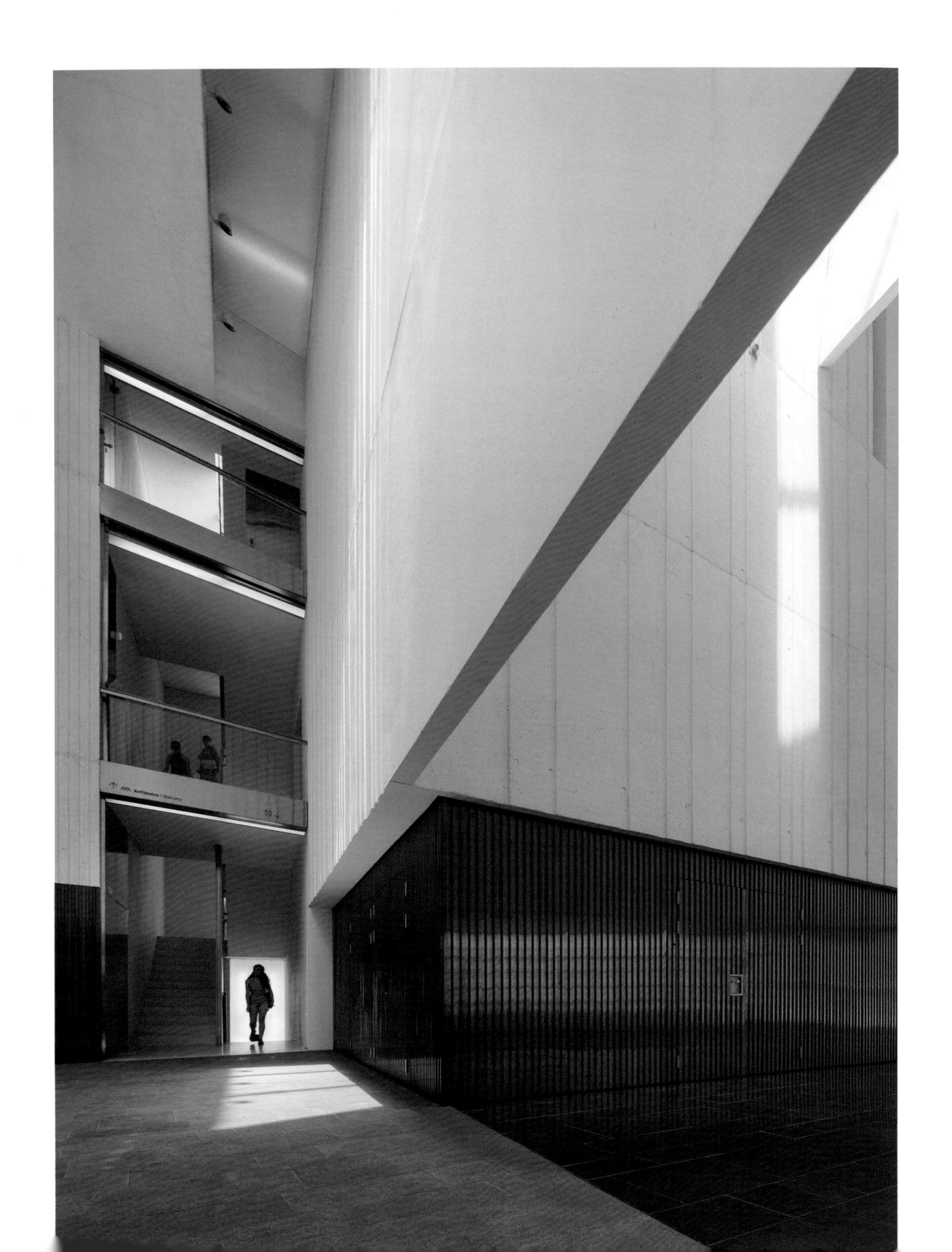
Anfiteatro | Balcony

Centro Federico Garcia Lorca

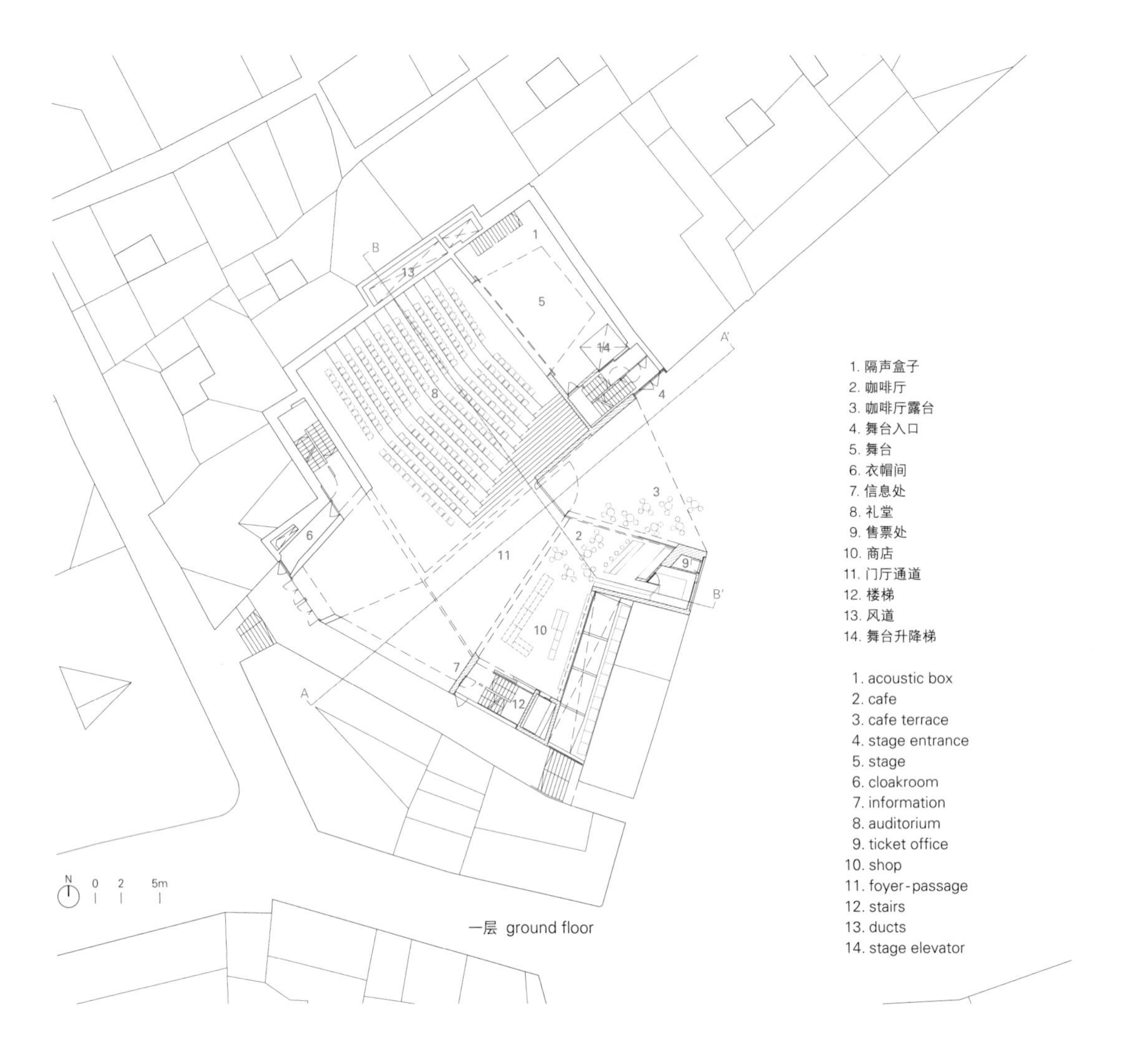

一层 ground floor

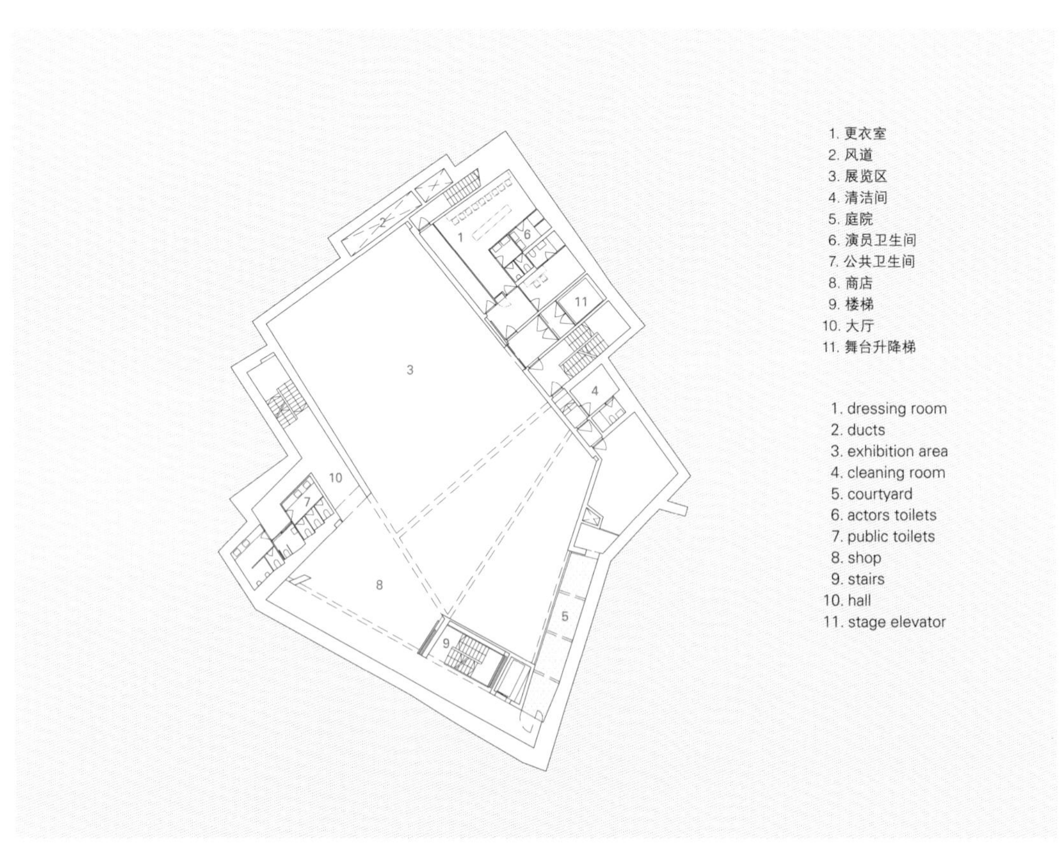

地下一层 first floor below ground

1. 办公室 1. offices
2. 会议室 2. meeting room
3. 档案室 3. archive
4. 厨房 4. kitchen
5. 清洁间 5. cleaning room
6. 卫生间 6. toilets
7. 大厅 7. hall
8. 庭院 8. courtyard
9. 机械间 9. machine room
10. 阳台 10. balcony
11. 风道 11. ducts

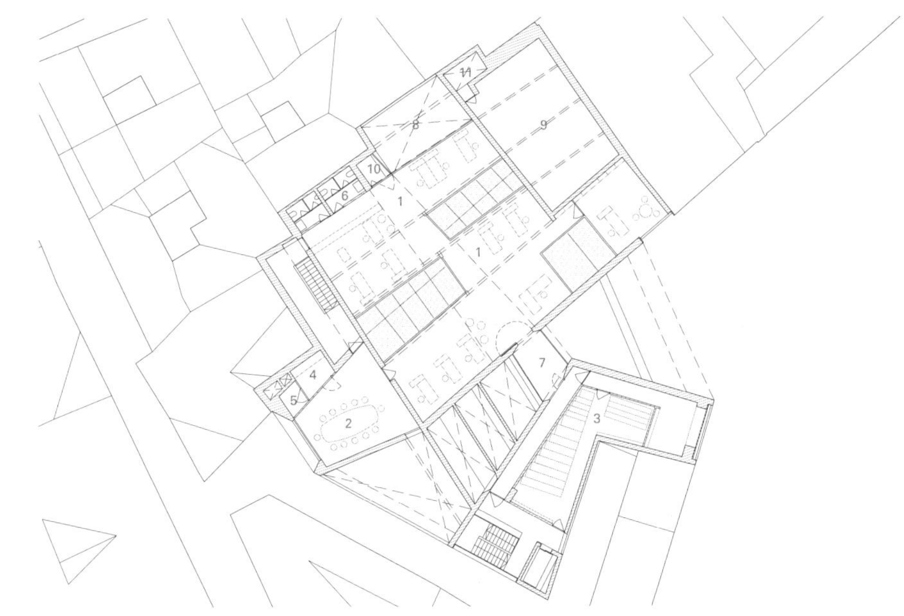

四层 third floor

1. 音乐舞蹈教室 1. music and dance room
2. 艺术室 2. art room
3. 视听室 3. audiovisual room
4. 迷你剧院 4. theater room
5. 社交空间 5. social area
6. 档案室 6. archive
7. 大厅 7. hall
8. 卫生间 8. toilets
9. 清洁间 9. cleaning room
10. 楼梯 10. stairs
11. 风道 11. ducts
12. 庭院 12. courtyard

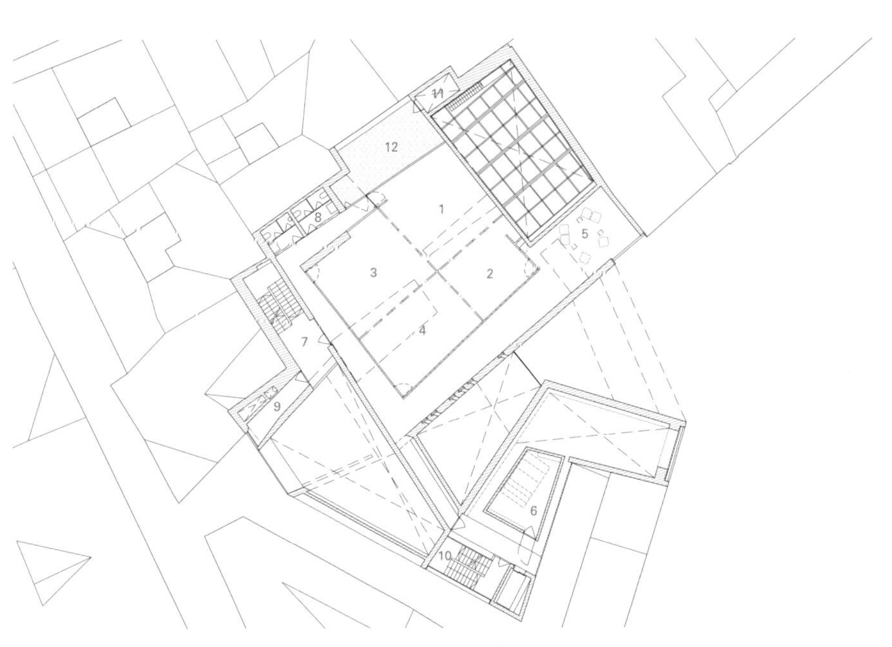

三层 second floor

1. 图书馆 1. library
2. 阅览室 2. reading room
3. 控制室 3. control
4. 舞台 4. stage
5. 画廊座位 5. seat gallery
6. 大厅 6. hall
7. 楼梯 7. stairs
8. 风道 8. ducts
9. 声控室 9. sound control room

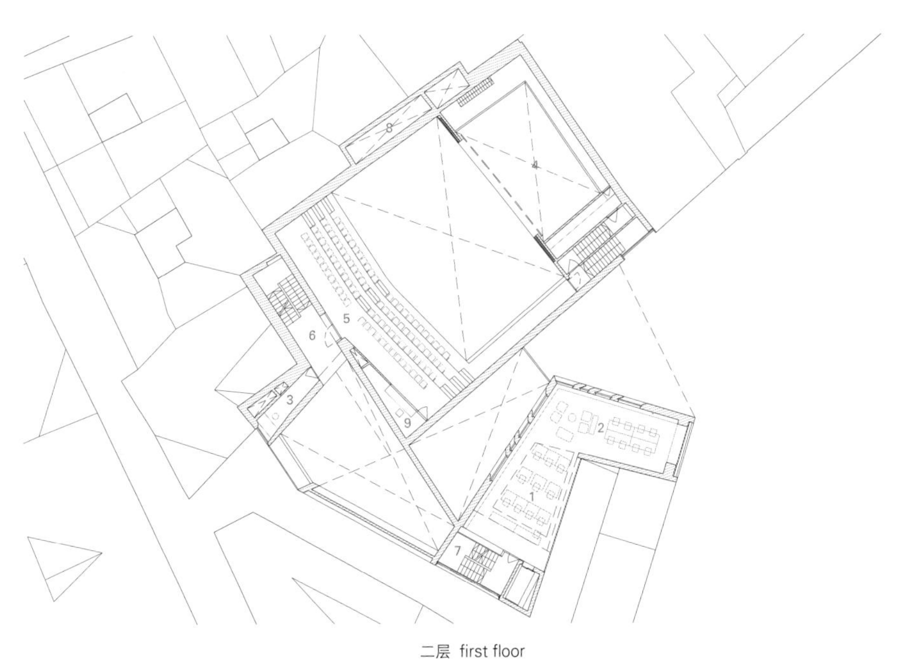

二层 first floor

dio, the Barcelona-based practice, for its sensitivity towards conserving the legacy and identity of the place.
The facility emerges as a new cultural landmark that blends naturally and fluidly into the historic fabric of the city of Granada, acquiring a significant urban representativeness.
The entrance to the new space is designed as a great threshold – a doorway that symbolizes the opening up of the Federico García Lorca Center, to the city, and vice versa. Its architecture incorporates the foyer and the public elements of the brief into the outdoor spaces to make up the setting.
The building forms a natural part of the public space, blends into its urban setting and creates a sequence of urban scenarios which allow the plaza in which it is situated to continue up to and into the building.
The boundary between the historic urban context and the architectural space is blurred by extending the entrance – or threshold – towards the public space of Plaza de La Romanilla.
The design strategy consists of sculpting the empty space into a great doorway at different scales. In this way, visitors enter via a passageway covered by beams and bridges which generate an interplay of shadows, creating new

three-dimensional and spatial relations.

In turn, the intervention naturally and fluidly integrates the plaza with the ground floor of the center, a space free of structural elements which accommodates a flexible, versatile program of activity.

The building, comprising two volumes of 4,700m², is laid out in a succession of different flexible and expressive spaces that house the various cultural uses. A 400-seat theater is equipped with an acoustic shell and mobile walls that soundproof it from the foyer and provide the necessary sound quality for concerts and plays.

A library, made up of two spaces of differing heights for the book stacks and the reading room, offers views of the suspended box that is the iron-clad archive where the poet's original manuscripts are safeguarded.

A gallery of generous dimensions occupies the basement, as well as a cafeteria, shop, offices, storage space and services.

The new center is built of in situ concrete, giving it a surface that is smooth to the eye and the touch. The texture, similar to natural stone, is in dialogue with the architecture of the historic center of Granada and allows the center to make its presence felt as a cultural landmark in the community.

项目名称：Federico García Lorca Center in Granada / 地点：Granada, Spain / 建筑师：Mara Partida, Héctor Mendoza–Mendoza Partida; Mónica Juvera, Boris Bezan–BAX studio (before MX_SI) / 合作方：Rodrigo Escamilla, Samuel Arriola, Oscar Espinoza, Albert Manubens / 施工技术员：Antonio Navarro Suarez / 项目经理：Artelia (Coteba), Marta Alda 结构工程师：B.O.M.A. Augustion Obiol, Diego Martin Saiz, Pere Vidal / 设备工程师：PGI Group, Joan Escanellas, Teresa Curiel / 消防顾问：Francesc Labastida / 声效工程师：Arau Acústica / 剧院设计顾问：Otto Seix, David Pujol / 客户：Consorcio Centro Federico García Lorca / 建筑面积：4,700m² / 造价：EUR 22,000,000 / 设计时间：2005 竣工时间：2017 / 摄影师：©Pedro Pegenaute (courtesy of the architect)

西里西亚大学新建的广播电视系大楼坐落在波兰南部卡托维兹一片几乎空置的土地上。工地上只剩下一栋废弃的建筑，最初的想法是要对其进行拆除的。

这座建筑的设计策略不是要设计一座标志性的新建筑，而在于建造和修复这座城市的一部分。为了实现这一点，建筑师感到有必要观察现有的大环境，发现它的独特之处，给它恢复一种特殊的建筑氛围和建筑个性。

新设计保留了现有的已经被废弃的建筑，在保护其特征的同时增加了一个扩建结构；它还包括一座高度较低的建筑，该建筑占据了内部街区区域，它与作为干预项目的重要部分的中央庭院产生了一种联系。

该设计的目的是要对现有建筑的审美产生敏感度，通过在建筑的上部建造一个由格子砖结构组成的抽象体量来展现它的材料性和视觉价值，这个格子砖结构延续了附近建筑的截面形式。根据乔纳森·塞尔吉森的说法："……从更大程度上对一种其他文化进行批判性的评价是可能的。作为一个局外人，当看到一些'其他的'东西，而不是通过文化渗透而变得熟悉时，则有可能看事情更加清楚……"

在外人充满新奇的注视下，建筑师觉得能够更好地欣赏这个地

卡托维兹西里西亚大学广播电视系

Silesia University's Radio and TV Department in Katowice

BAAS Arquitectura

方——建筑的美丽、实体结构与架空空间的比例、历史悠久的用黑砖砌成的天井、有特色的宿舍以及逐渐破败的墙壁的纹理。

通过这个过程，他们可以找到失去的部分，而这正是可以最大限度地发挥地方特色的部分，并没有将其转化为其他事物的想法，同时也能理解并尊重先前的存在并保持它的完整性。在这个方法中，建筑设计居于第二位，由新建筑自己自然地嵌入。

新建筑完善了街区，并遵循街道的走向，与原有建筑的形状和高度相融合。外部饰面和颜色来自于竞赛最初阶段提议拆除的不显眼的现有建筑。

这种陶瓷材质得以侵入室内空间，提供独特的氛围和照明效果。建筑的露台旨在增强建筑的现有外观存在感，将纹理的特质带到建筑的内部。沿着内庭院的一条道路将提供居住空间，并使得这些设计策略具有公共性和可利用性。

一段笔直的楼梯沿着整座建筑一直通向顶部，为学生们在下课之后提供了一个聚会的空间；建筑的透明度使他们能够从内部庭院观察到外部的运动，就像电影中的临时演员一样。

这座新大楼将整个地块填满，同时又在中间空出了一个中央庭院，这成为这所大学院系的新建演播室和演讲室周围所有社交活动的关键元素。这就是嵌入的有机特性，到此一游的游客可能会问自己：这里一直都是这样的吗？

©Jesus Arenas

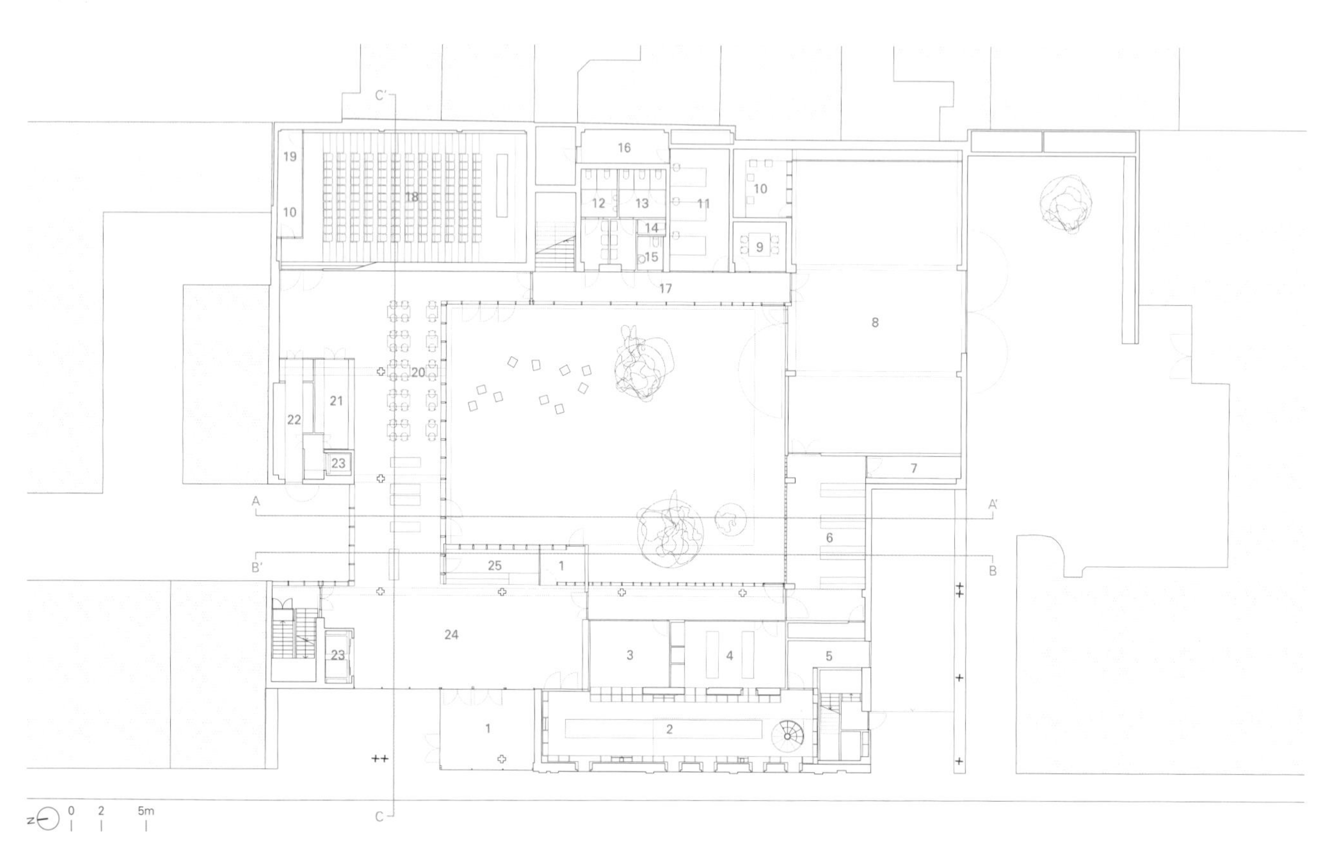

1. 游廊 2. 图书馆 3. 色彩实验室 4. 照明设备库房 5. 图书管理员办公室 6. 道具和杂志 7. 电影设备库房 8. 电影工作室 9. 声音实验室 10. 导演室 11. 剪辑室 12. 男士卫生间 13. 女士卫生间 14. 无尘室 15. 特殊卫生间 16. 库房 17. 交流区 18. 放映室 19. 译制室 20. 学生俱乐部 21. 吧台 22. 厨房 23. 电梯 24. 大厅 25. 接待处

1. vestibule 2. library 3. color laboratory 4. light storeroom 5. librarian's office 6. props and magazine 7. storeroom for film equipment 8. movie studio 9. sound laboratory 10. directing room 11. editing room 12. men's toilet 13. women's toilet 14. clean room 15. adapted toilet 16. storeroom 17. communication area 18. cinema room 19. translation and interpretation room 20. club for students 21. bar 22. kitchen 23. lift 24. hall 25. reception

一层 ground floor

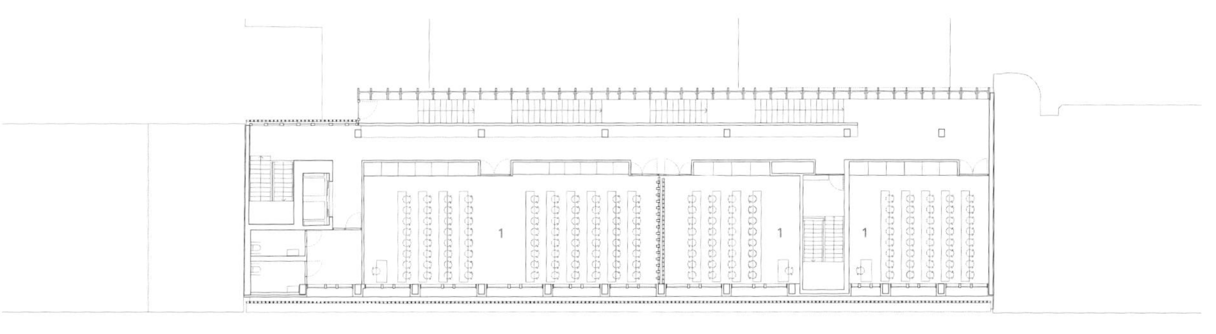

1. 大教室 1. lecture room
四层 third floor

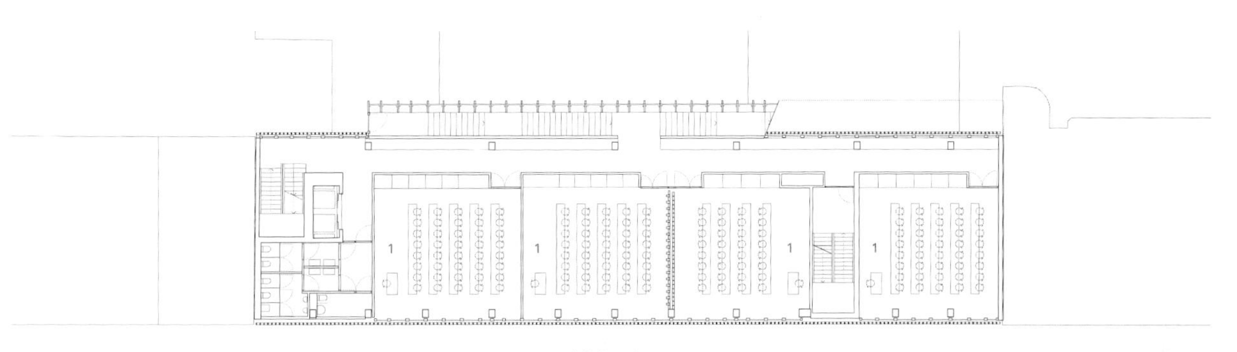

1. 大教室 1. lecture room
三层 second floor

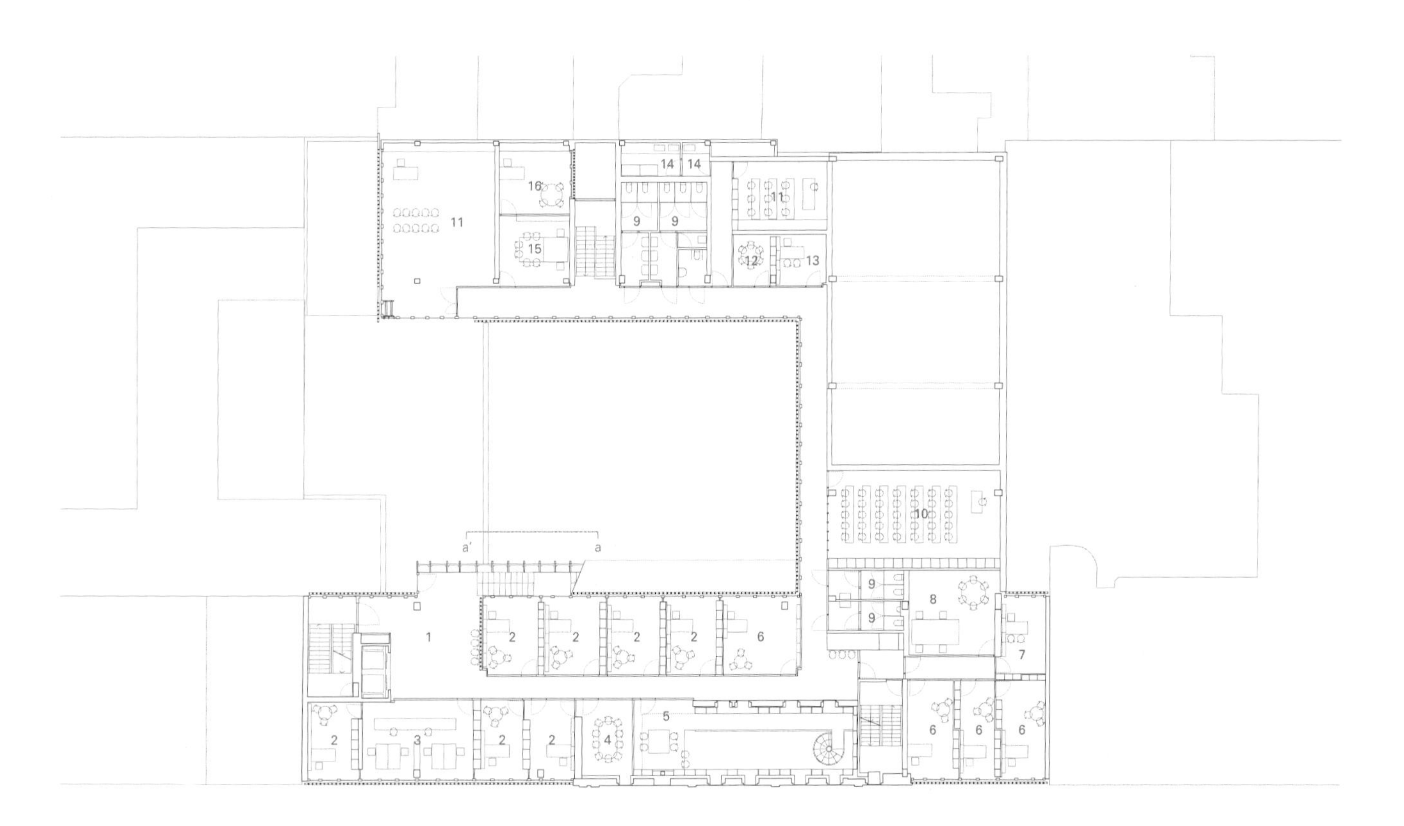

1. 走廊 2. 办公室 3. 餐厅 4. 会议室 5. 图书馆 6. 行政区 7. 图像处理系 8. 工人房 9. 卫生间 10. 40人大教室 11. 摄影室 12. 学生教室 13. 游廊 14. 暗室 15. 制片系 16. 计划经理办公室
1. hallway 2. office 3. dining room 4. conference room 5. library 6. administration 7. department of image implementation 8. room for workers 9. toilet 10. lecture room for 40 people 11. photographic studio 12. student room 13. vesibule 14. dark-room 15. department of production 16. plan manager's room
二层 first floor

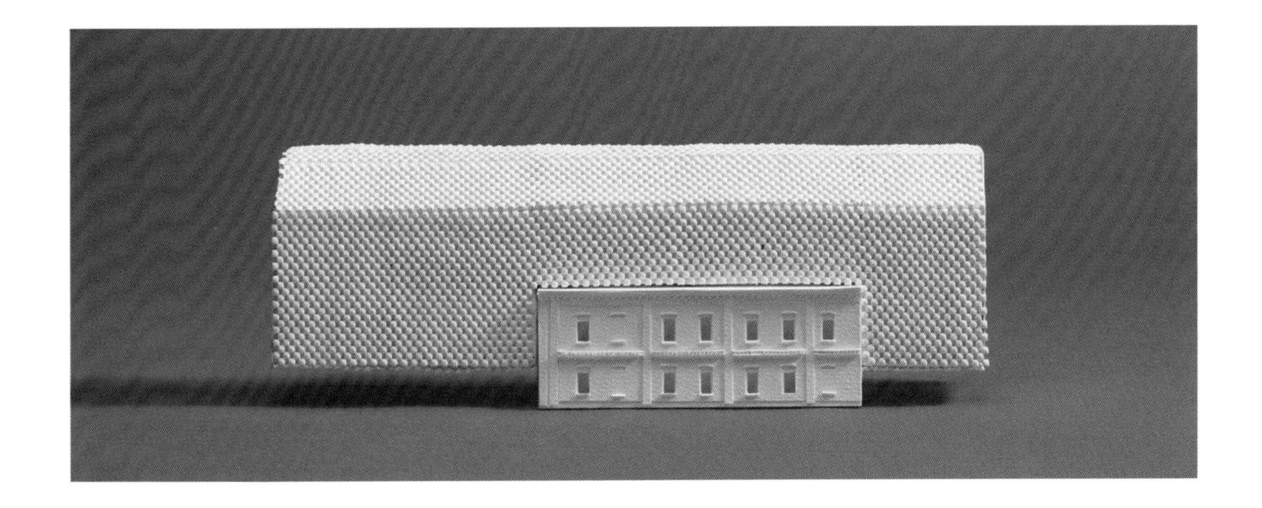

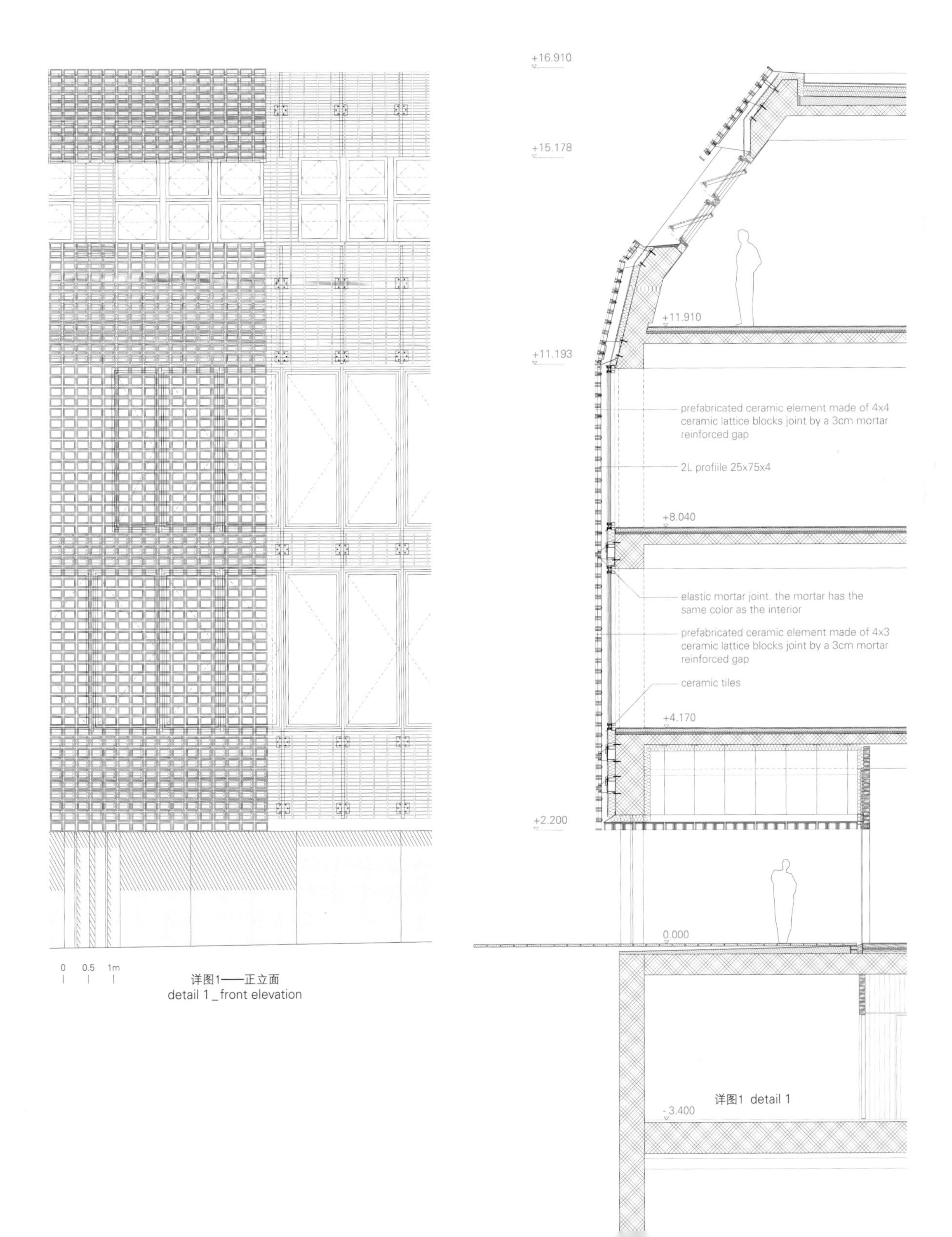

详图1——正立面
detail 1_front elevation

详图1 detail 1

The Radio and TV department of the new Silesia University is inserted into a consolidated area of Katowice on an almost vacant plot. An abandoned building – initially earmarked for demolition – was all that was left on the site. The design strategy for the building consisted not in designing an iconic new building, but building up and mending an existing piece of the city. To achieve this, the architects felt the need to observe the existing context, discover what made it unique, and restore to it a special atmosphere and personality.

The new design preserves the existing abandoned building, adding an extension to it while protecting its character; it also includes a building of lower height, occupying the interior block area, which confers to the central courtyard, the prominence of the intervention.

The design aims to be sensitive to the existing building aesthetics, taking advantage of its materiality and visual values by building on top of it, an abstract volume made out of a brick latticework, which follows the neighboring section. According to Jonathan Sergison: "...a greater degree of

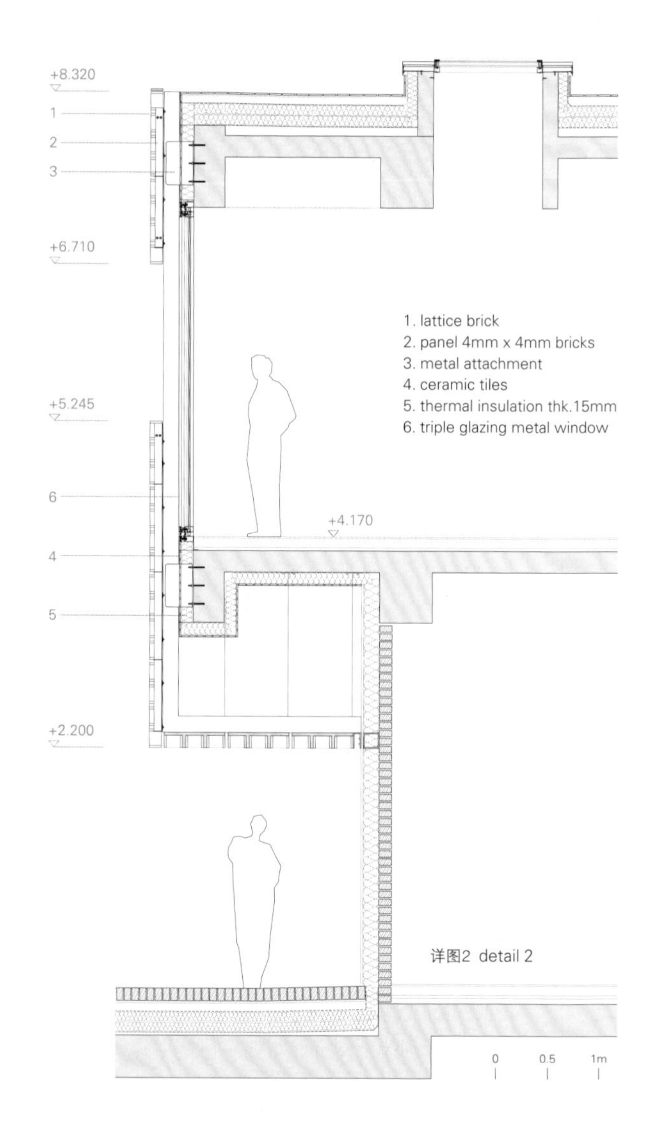

详图2 detail 2

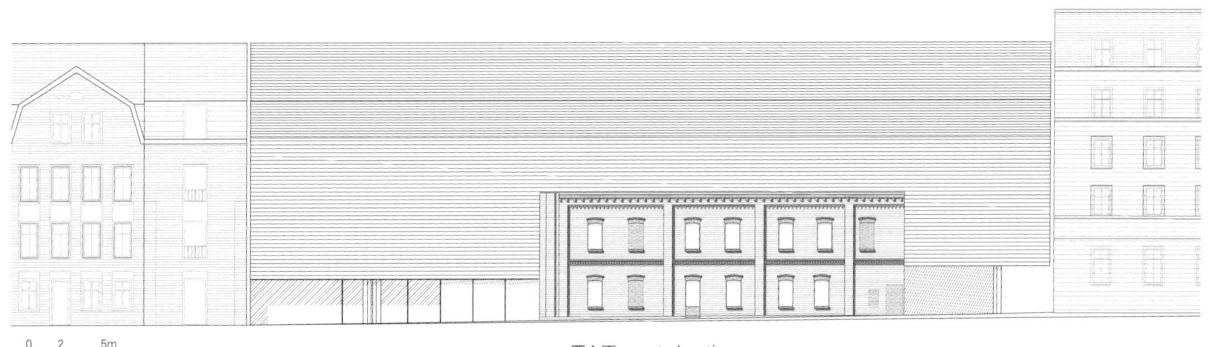

西立面 west elevation

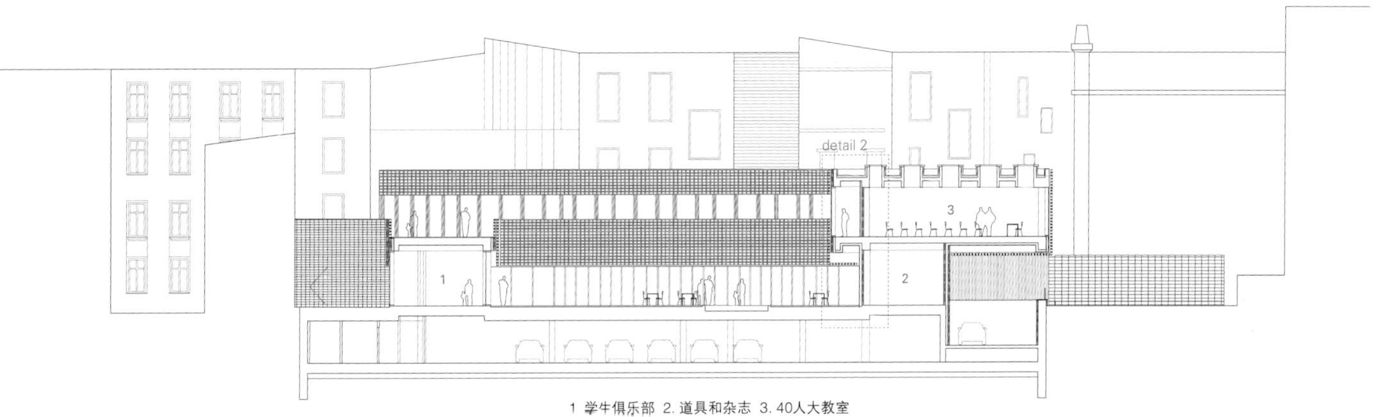

1 学生俱乐部 2. 道具和杂志 3. 40人大教室
1. club for students 2. props and magazine 3 lecture room for 40 people
A-A' 剖面图 section A-A'

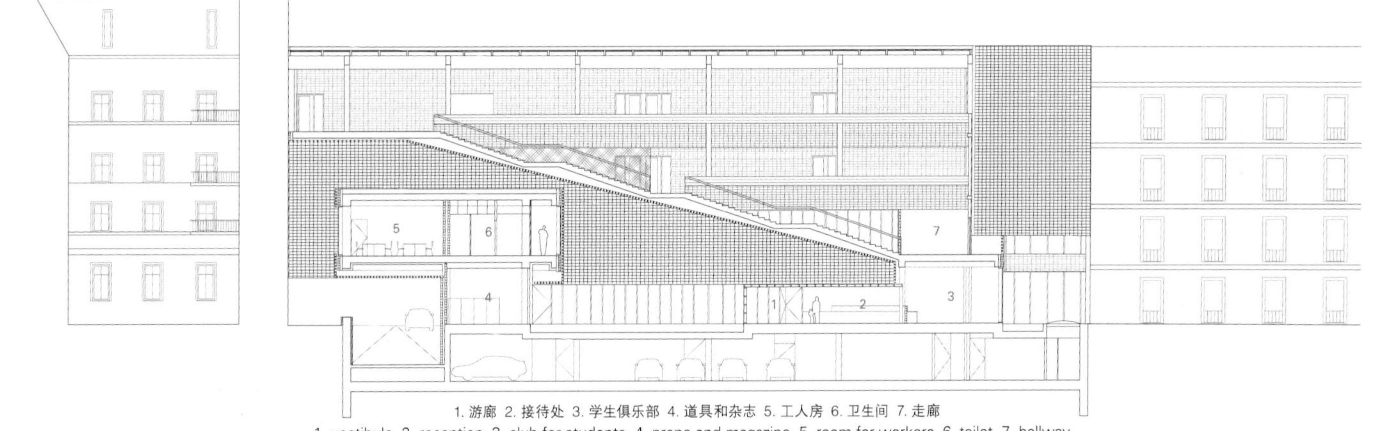

1. 游廊 2. 接待处 3. 学生俱乐部 4. 道具和杂志 5. 工人房 6. 卫生间 7. 走廊
1. vestibule 2. reception 3. club for students 4. props and magazine 5. room for workers 6. toilet 7. hallway
B-B' 剖面图 section B-B'

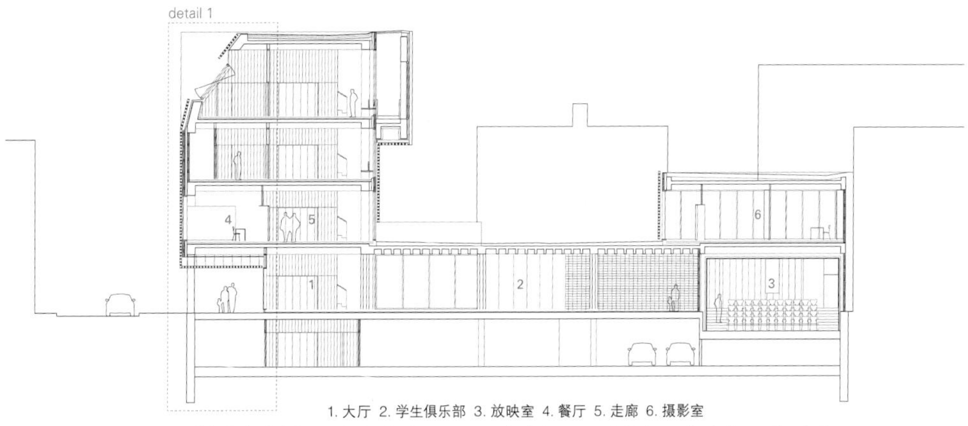

1. 大厅 2. 学生俱乐部 3. 放映室 4. 餐厅 5. 走廊 6. 摄影室
1. hall 2. club for students 3. cinema room 4. dining room 5. hallway 6. photographic studio
C-C' 剖面图 section C-C'

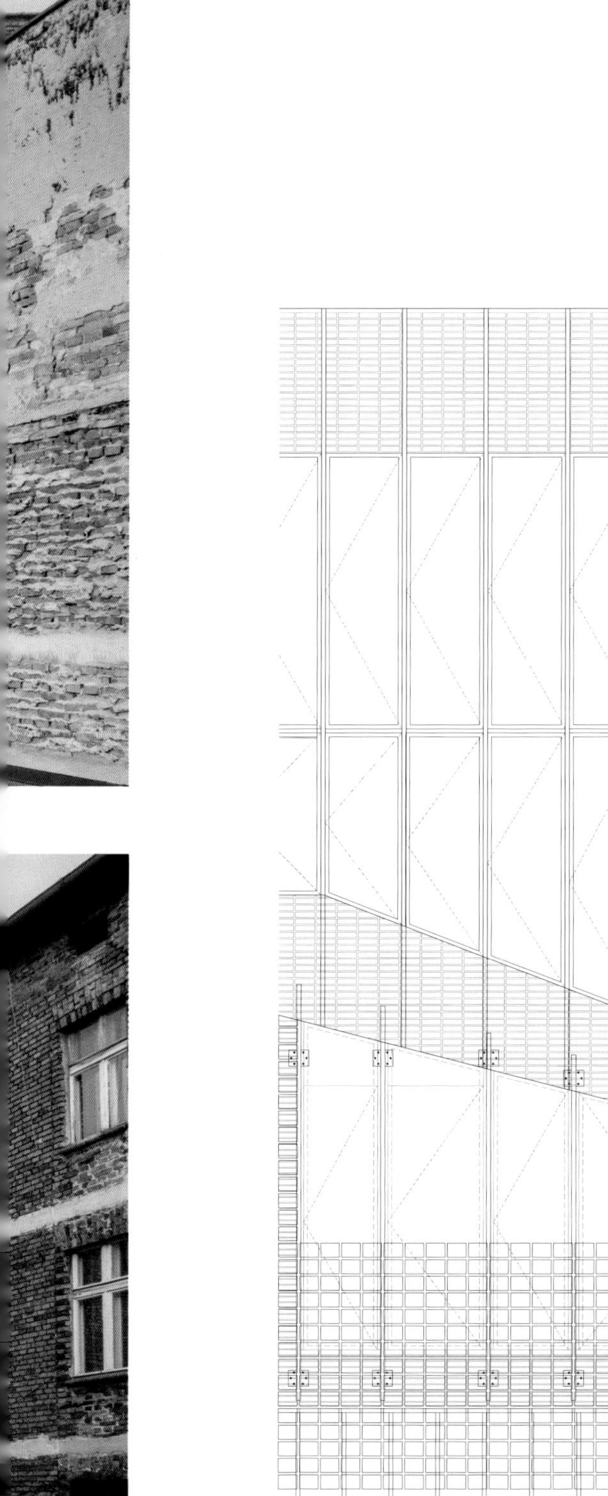
a–a' 详图 detail a-a'

critical appraisal is possible of a culture that is not one's own. As an outsider it is possible to see things with more clarity when looking at something that is "other" and not rendered familiar through cultural immersion...".
With the fresh eyes of outsiders, the architects felt better able to appreciate the place – the beauty of the building, the proportion between mass and void, the patina of a darkened brick over the years, the characteristic dormers and the texture of the gradually dilapidating walls.
Through this process, they could find the missing piece, the one that maximizes the character of the place, without the ambition of transforming it into something else, and at the same time understand and show respect for the preexisting and complete it. This is an approach in which architecture becomes secondary and decides to naturally insert itself.
The new building completes the block and follows the street alignment, merging with the preexisting shape and height. The exterior finish and color are taken from the inconspicuous existing building that the competition initially proposed to demolish.
This ceramic texture invades the interior spaces to give a unique atmosphere and lighting. The block patio aims to empower the existing exterior of the building, bringing the qualities of the texture to the interior of the block. A path along the interior patio will make for habitable spaces, and will allow for these strategies to become public and available.
A linear staircase follows the entire height of the building, generating a space for students to gather when finishing classes; the transparency of the building allows them to observe movement from the interior patio, like extras in a movie.
The new building fills up the whole plot whilst at the same time hollows out a central courtyard, which becomes the key element for all the social activities taking place around the studios and lecture rooms of this new university department. Such is the organic nature of the insertion that a visitor to this place may well ask themselves: Has this always been here?

项目名称：Silesia University's Radio and TV Department, Katowice / 地点：Katowice, Poland / 建筑师：BAAS arquitectura
当地合伙人：Grupa 5 Architekci, Maleccy Biuro Projektowe / 项目建筑师：Jordi Badia, Rafał Zelent, Wojciech Malecki
项目团队：Alba Azuara, Jordi Framis, Mireia Monràs, Daniel Guerra, Raül Avilla, Joan Ramon Pastor, Cristina Luis, Mariona Guàrdia, Rafael Berengena, Xavier Gracia, Enric Navarro, Gonzalo Heredia, Carla Llaudó, Eva Olavarria / 客户：University of Silesia in Katowice (WRiTV) / 总建筑面积：5,390m² / 竣工时间：2017 / 摄影师：©Adrià Goula (courtesy of the architect)

颂扬多元化

n Praise o
School Archite

当今学校建筑

我们的学习时光是我们生命中情感最丰富的时期，我们的童年中有很大一部分是在学校度过的。根据意大利教育改革家洛里斯·马拉古齐所说，孩子们主要通过互动学习的方式获得知识，而且他们身边有三位老师：周围的成年人、同龄人以及他们周围的环境。

由此显而易见，学校的建成环境对我们的孩子的教育和他们的情感和心理健康起着至关重要的作用。建筑师需要了解什么样的环境有助于学习。我们如何创造让儿童和教师认同的场所，并营造一种信任和开放的学习氛围？什么样的形式，什么样的光线和通风，什么样的户外空间最有利于孩子们，使他们感到安全

Our school years are the most emotionally charged periods of our lives and a significant proportion of our childhood is spent in school. According to Italian educational reformer, Loris Malaguzzi, children primarily learn through interaction and have three teachers: the adults around them, their peers, and their immediate environment.

It is obvious therefore that the built environment of schools plays a critical role in the education of our children and in their emotional and psychological well being. Architects need to inform themselves as to what kind of environments are conducive to learning. How do we create places that children and teachers identify with and that foster an atmosphere of trust and openness to learning? What sort of forms, what sort of light and ventilation, what kind of outdoor spaces benefit children

南港学校_South Harbor School/JJW Arkitekter
圣伊西多尔学校集团扩建工程_Saint Isidore School Group Extension/ANMA
Žnjan-Pazdigrad小学_Žnjan - Pazdigrad Primary School/x3m
TTC槟椥精英幼儿园_TTC Elite Ben Tre Kindergarten/KIENTRUC O
Nová Ruda幼儿园_Kindergarten Nová Ruda/Petr Stolín Architekt

颂扬多元化——当今学校建筑_In Praise of Diversity - School Architecture Today/Anna Roos

f Diversity

cture Today

感并在心理上做好学习的准备？正如温斯顿·丘吉尔的那句名言："我们塑造了我们的建筑，而建筑反过来也塑造了我们。"因此，我们长期以来的学习环境也将深刻地影响我们的孩子，并影响他们的学习时光，这是他们未来获得成功的成人教育和未来职业前景的重要基石。

本书所展示的项目展示了全球各地学校设计的解决方案，从哥本哈根的一所具有阶梯式外观的学校，到尼斯的一所带有亭子、遮篷结构的学校，斯普利特的一所由光滑混凝土建成的学校，越南一座嵌入山脚下的幼儿园，以及最后提到的捷克共和国的一所采用玻璃纤维覆盖的幼儿园。

the most and make them feel safe and psychologically ready to learn? As Winston Churchill famously said, "We give shape to our buildings and they, in turn, shape us." So the environment that we spend our years of education will shape our children profoundly and affect their school careers, a crucial stepping-stone to their successful adult education and their future career prospects.
The projects shown in this book showcase the variety of solutions to school design around the globe, from a cascading school in Copenhagen to a village of pavilion sheds in Nice, a sleek concrete school in Split, a kindergarten embedded beneath a hill in Vietnam, and finally a glass-fiber clad kindergarten in the Czech Republic.

颂扬多元化——当今学校建筑
In Praise of Diversity - School Architecture Today

Anna Roos

在历史上，社会变革、新教育理论和开创性的学校建筑彼此是相互影响的。尽管如此，建筑师也应该注意到，学校建筑在教学方法上比时尚更有生命力，所以学校建筑应该提供可以灵活使用的空间。这里需要有各种各样的空间，如同一座城市的缩影，街道、广场和封闭的空间为自由活动、聚会、交流、单独工作或分组工作提供了空间，从而让人们感觉到受到了保护或者受到了挑战。我们应该抛弃房间只有一种功能的观念。教室应鼓励不同的课桌配置，而不是只具备单一的功能，它们应该考虑到智能组合和空间连接，毕竟正面教学只是一种选择。

将现代学校建筑简化为一套统一的元素并仅仅作为学校的用途来使用，在今天已经是不可能的了。新式的空间和原始类型的空间提供了多元化、具有启发性的学习环境，也可以开放供当地社区使用。哥本哈根海港区的南港学校（第156页）就是一个很好的例子，它被设计成一个巨大的城市景观，有室外阶梯，可以让人直接从外头前往升高的操场。由于该地区缺乏一个具有连贯性的焦点，因此学校的一部分设计要求是作为当地居民的聚会场所和文化聚集点。JJW建筑师事务所的目标是“邀请城市进入学校，并将学校与城市联系起来”。因此，这所学校被设计成了一个U形的结构，围绕着一个向城市开放的种植操场庭院展开。JJW

Historically, social change, new education theories and pioneering school architecture have mutually influenced one another. Having said that, architects should be mindful that school buildings outlive fads in teaching methods, so school buildings should provide scope for flexible use. There needs to be a variety of spaces, like a city microcosm, with streets, squares and enclosed spaces providing room to move around freely, gather together, converse, work alone or in groups, feel protected or feel challenged. We should move away from the notion that a room has only one function. Classrooms should allow for different configurations of desks, rather than being mono-functional, they should allow for the intelligent combinations and spatial connections, after all frontal teaching is only one option.

It is no longer possible to reduce modern school architecture to a set of uniform elements that serve only as a school. New and original types of spaces provide varied and inspiring learning environments that are also open for use by the local community. South Harbor School in Copenhagen's harbor precinct (p.156) is an excellent example of a school that has been designed as an enormous urban landscape with cascades of stairs that rise up to connect with raised playgrounds. As the area lacked a coherent focus, part of the brief was for the school to also function as a meeting place for local residents and as a cultural gathering point. JJW Arkitekters aim was to "invite the city into the school and connect the school with the city". The school has therefore been designed on a U-shaped plan enclosing a planted playground courtyard that opens to the city. JJW see this as a "combination of school yard and city square". Generous cascading timber stairs connect the school and the harbor and can be used for teaching, a recreation area for pupils, or a meeting place for locals. By giving the

©Torben Eskerod
南港学校，丹麦 South Harbor School, Denmark

建筑师事务所将它视为"校园和城市广场的结合体"。宽敞的木楼梯连接着学校和港口，可用于教学、学生娱乐区或当地人的聚会场所。通过赋予楼梯额外的连接方式，从街道层面到学校的通道变得更加流畅，通过使学校的外围尽可能吸引人，使各元素之间的空间更适合居住，建筑师为建筑增添了更多的价值。从公共空间、半公共空间到更为私密的空间，它们之间层次分明，而阶梯式的室外空间则被设计成额外的学习、娱乐和活动空间。各种不同规模、不同类型的空间被创造出来，以满足这一需求：在两个教室轮流进行学习活动的小学生群体能够参与到学校的集会中来。

位于尼斯的圣伊西多尔学校（第172页）由ANMA建筑事务所设计，是一组木结构的亭子。与之前在哥本哈根的项目一样，它也有一个升高的户外运动场，在那里可以看到周围的山丘。学校被设计成带有防风效果的操场，操场有升高的木质露台和新种植的树木。学校南部有一个树木林立的公园，为这里的交通创造了一片绿色缓冲带。建筑师营造了一种有韵律感的连体坡屋顶亭子，缩小了学校校园的规模，使它更适合于儿童的成长。明亮的红色和黄色风琴百叶窗强调了学校的趣味性。

克罗地亚的Žnjan-Pazdigrad小学（第184页）是位于斯普利特近郊的一个相对较新的郊区，它不仅被设计成一所学校，还被设计成社区的集体空间，以前这里可是非常缺乏公共空间的。尽管实际上，场地太小了，不足以容纳设计要求中规定的空间，但是建

stair extra articulation the access from the street level to the school is made more fluent and by making the periphery of the school as inviting as possible and by making the space between elements more habitable, the architects have added value. A hierarchy has been established from public to semi-public and more private spaces, while the terraced outdoor spaces have been designed as additional learning, recreational and activity spaces. A myriad of different kinds of spaces with varied scales have been created that cater for the entire school assembly to intimate groups of children having a learning session between two classrooms.

Saint Isidore School (p.172) in Nice, designed by ANMA is a cluster of timber pavilions, which, like the previous project in Copenhagen, also has a raised outdoor playground with views to the surrounding hills. The school has been designed around a wind-shielded playground with raised timber terraces and newly planted trees. A tree-filled park to the south creates a green buffer to the traffic. By creating a rhythm of conjoined pitch-roofed pavilions, the architects have reduced the scale of the school campus and have made it more child-friendly. Bright red and yellow concertina shutters emphasis the playful nature of the school.

Žnjan-Pazdigrad Primary School (p.184) in Croatia is a relatively new suburb on the outskirts of Split which has been designed to not only function as a school, but also as a collective space for the neighborhood, which previously lacked communal public spaces. Although the site was actually too tight to accommodate the brief, the architects maximized the terrain by creating a new topography on the sloping site. A series of stacked decks intertwine sports facilities and classroom spaces with multipurpose spaces. The clean-cut, refined expression of slabs and structure of criss-cross concrete columns and the large-scale typography painted on the facades of one of the embedded volumes are the only vaguely decorative elements. Projecting floor slabs

圣伊西多尔学校集团扩建工程，法国
Saint Isidore School Group Extension, France

Žnjan-Pazdigrad小学，克罗地亚
Žnjan-Pazdigrad Primary School, Croatia

筑师仍然想办法在斜坡上创建了一种新的地形，使该区域实现面积最大化。一系列堆放的平台将体育设施和教室空间与多用途空间交织在一起。简洁、精致的石板和纵横交错的混凝土柱结构，以及绘制于其中一个嵌入式体量的立面上的大型字体，是仅有的模糊的装饰元素。突出的地板使教室免受太阳光的直接照射。通过将建筑结构暴露出来，建筑师创造了一种干净整洁的表达方式。与本系列中所提到的其他学校一样，该学校也有升高的操场、体育设施和多用途空间，同时它们也可供当地社区使用。吸声天花板的使用能够吸收声音，减少噪声的压力，让学生更好地集中注意力。

越南的TTC槟椥精英幼儿园（第202页）就是一个很好的适合儿童使用的幼儿园案例。就像霍比特人的房子一样，KIENTRUC O建筑事务所把教室嵌入一座顶部有一个光滑的矩形玻璃盒的绿色小山下面。旋转的楼梯和点缀着绿色植物的大面积木质地板，模糊了内部和外部空间之间的边界。建筑师没有选择设计一个个独立的教室，而是设计了可以以多种方式来使用的流动性空间。开放式的平面布局实现了热带气候下的良好的空气流通和冷却效果。通过将人与自然融为一体，建筑师的目标是创造这样一个场所，孩子们可以“在一个能激发他们最大潜能的有趣环境中表达、探索和学习”。

Nová Ruda幼儿园（第216页）位于捷克共和国利贝雷茨市近郊的Vratislavice nad Nisou区。Petr Stolín建筑师事务所专注于创建各种特殊的结构，通过这些结构，孩子们可以探索幼儿园的奥秘。设计的目的是充分利用斜坡地段的条件，这里最初似乎并不

shade classrooms from direct sun. By exposing the structure, the architects have created a clean-cut unfussy expression. Like other schools in this series, the school has elevated playgrounds, sports facilities and multi-purpose spaces that can also be used by the local community. Acoustic ceilings absorb sound, reducing the stress of noise, allowing pupils to concentrate better.

A wonderful example of a child-friendly kindergarten is the TTC Elite Ben Tre Kindergarten (p.202) in Vietnam. Like a Hobbit's house, KIENTRUC O has embedded the classrooms beneath a green hill surmounted by a sleek, rectangular glass box. Swirling staircases and sweeping timber floors interspersed with greenery blur the boundary between interior and exterior spaces. Instead of designing separate classrooms, flowing spaces can be used in a versatile manner. The open plan layout allows for good airflow ventilation and cooling of the spaces in the tropical climate. By integrating people and nature, the architects' aim was to create a place where children can "express, explore and learn in a playful environment that brings out the best in them".

Kindergarten Nová Ruda (p.216) is located in Vratislavice nad Nisou, on the outskirts of Liberec, Czech Republic. Petr Stolín Architekt focused on creating various special configurations through which children can discover the kindergarten. The design resulted from wanting to utilize the conditions of a sloping site that initially did not seem ideal for a kindergarten. Each room has a different function, such as a playroom, a classroom, and a sleeping area, so that children can walk up and down inside the kindergarten as if they were exploring a big jungle. In between the two main bodies of the kindergarten, terraces, gardens, and green space with various play elements are created, allowing children to play inside and outside. Meanwhile, the materials used in the kindergarten are mainly wood and fiberglass which bring a feeling of intimacy and warmth.

TTC槟椥精英幼儿园，越南
TTC Elite Ben Tre Kindergarten, Vietnam

适合用来建造幼儿园。每个房间都有不同的功能，例如，游戏室、教室和睡眠区，这样孩子们就可以在幼儿园里来回走动，就像在探索一个大丛林一样。在幼儿园的两座主体建筑之间，建筑师创造了带有各种游戏元素的露台、花园和绿地，允许孩子们在室内和室外环境玩耍。同时，幼儿园使用的材料主要是木头和玻璃，给孩子们带来一种亲密和温暖的感觉。

幼儿园是孩子们除了他们自己的家之外在一天里度过的时间最多的地方：建筑师希望孩子们的日常生活充满探索。

设计良好、受欢迎的学校更可能受到儿童的尊重，而不是单纯具有高科技的功能性建筑。孩子们需要认同他们的环境，从而增强他们与周围环境的情感联系。通过激活建筑中较小的元素，例如，提供窗台、架子和壁架，通过这些地方，可以让孩子们拥有展示手工艺品的地方，同时也让孩子们能够在其中找到家的感觉。事实证明，在充满压力的环境下，学习和吸收信息是困难的，因此，学校应该是培养孩子情感健康的安全场所，这一点至关重要。像木材这样的材料可以吸收声音，并有助于创造安静的空间，帮助孩子们集中注意力。建筑师必须预见到他们的用户的愿望，这意味着他们的设计要符合人体工程学，并且要非常细心和富有同情心。可以转化为有利于儿童的设计的、以儿童为中心的教育方法，应该能促进儿童友好型学校的建成。学校应该是改善生活状态的地方，在这里我们可以促进孩子整体的身心发展。

Kindergarten is the place where children spend most of their time, in a given day, except for their home: the architects wanted the children's daily lives to be full of exploration.

Well-designed, welcoming schools are more likely to be treated with respect by children than purely functional hi-tech buildings. Children need to identify with their environment thereby enhancing their emotional affinity with their surroundings. By activating the smaller elements of a building, like for instance providing cills, shelves and ledges that create opportunities to display handicrafts and enable children to appropriate a space making themselves feel at home in it. It has been proven that it is difficult to learn and absorb information under stressful circumstances, so it is crucial that schools should be safe places that nurture the emotional well being of our children. Materials like timber absorb sound and help to create acoustically calmer spaces, helping children to concentrate. Architects must anticipate the wishes of their users, which means designing ergonomically and with great care and empathy. A child-centered approach to education that translates into child-friendly design should foster child-friendly schools. Schools should be life-enhancing places that promote the development of the whole child, physically, mentally and emotionally.

1. Hertzberger, Hermann, Lessons for Students in Architecture, Rotterdam: 010, 1991.
2. Meuser, Natascha (ed.), Handbuch und Planungshilfe, Schulbauten, Berlin: DOM Publishers, 2014.
3. Müller, Thomas and Schneider Romana, The Classroom, from the late 19th century until the present day, Bonn: Wasmuth, 2010.

南港学校
South Harbor School

JJW Arkitekter

courtesy of the architect

哥本哈根港的一所新式学校

当JJW建筑师事务所在建筑设计竞赛中赢得南港学校的项目时，这个旧港区正在被开发成一个新的混合用途市区，但是该区缺乏一个可供集会的场所。此次设计竞赛的主旨正是要打造一个既能作为学校又能作为文化聚集点的项目。

JJW建筑师事务所的设计是邀请城市进入学校，并将学校与城市联系起来，通过将水域转化为额外的教室这种方式，进而达到充分利用港口位置的目的。同时，各种规模的不同类型空间也为学生创造了无数的聚会机会——从学校集会到两班同学之间的小型学习研讨。这所学校已获得多项大奖，包括2016年的WAN教育奖等。

学校和城市

这所学校被设计成拥有公共、半公共和更为私密空间的自然层次，而室外区域被设计成额外的学习、休闲和活动区域。

为了在学校、周边环境和港口之间建立一种交流，项目地面层的特色是混合了校园和城市广场。一个巨大的楼梯将学校与港口连接起来，用作额外的教室、学生的娱乐区以及哥本哈根市民生动的公共集会场所。

根据学生需要进行调整的空间

这所学校在一定程度上是按年龄划分的，低年级的在最底层，而高年级的在最上层。容纳大一点学生的上部楼层同时为年龄的划分和年龄的融合做好了准备。这样，内部和外部空间的设计就考虑到了学生在不同的人生阶段能够处理问题的程度，以及让他们拥有宾至如归和安全的感觉。为了给年幼的孩子们一种安全感，低年级的班级有他们自己的家庭区域，这一点把他们与年龄大的学生分开了。更多的高年级学生共享一个更大的、整合各个年龄组的家庭区域，分布在两个楼层上。在那里，学生们跨越了不同的年龄群体，而形成了更为庞大的社区的一部分。

鲜活立面

这座建筑的特点是有一个生动鲜活的立面，不仅创造出变化，而且支撑了建筑的教育功能。

这个充满活力的立面充当了一个高效、智能的建筑外围护结构。能够自动开启的窗户可控制夜间制冷，特别开发的减噪窗可实现自然通风，而建成的立面屏风既能防晒，又能在寒冷季节增加一个热缓冲区。建筑师采用了一种比较复杂的表面材料——预制压缩矿棉镶板，

并将它们用于一种强烈而又生动的表现形式。最外层是百分之百回收的铝薄板。根据观看者所处的位置不同，立面会发生有趣的变化。

持续性

学校遵循DGNB（德国可持续建筑委员会）方法的参数，该方法侧重于技术、经济和社会可持续性的综合考量。放学后，建筑的使用支持了这种可持续方法的社会层面，例如，室外区域同时可以作为校园和公园使用。

A New Profile School in the Harbor of Copenhagen

When JJW Arkitekter won the architectural competition for South Harbor School, the old harbor region was being developed into a new mixed-use quarter, but the district lacked an assembly area. The competition brief was to create a project that could function both as a school and a cultural gathering point.

JJW Arkitekter' design invites the city into the school and connects the school with the city, utilizing the harbor location by transforming the water into an extra classroom. At the same time, varied spaces of different scales create numerous gathering opportunities for students – from school assemblies to smaller learning sessions between two classmates. The school has received several awards including the WAN Education Award 2016.

The School and the City

The school is designed with a natural hierarchy of public, semi-public and more private spaces, while the outdoor areas are designed to be used as additional learning, recreational and activity areas.

To create an exchange among the school, the neighborhood and the harbor, the ground floor features as a hybrid

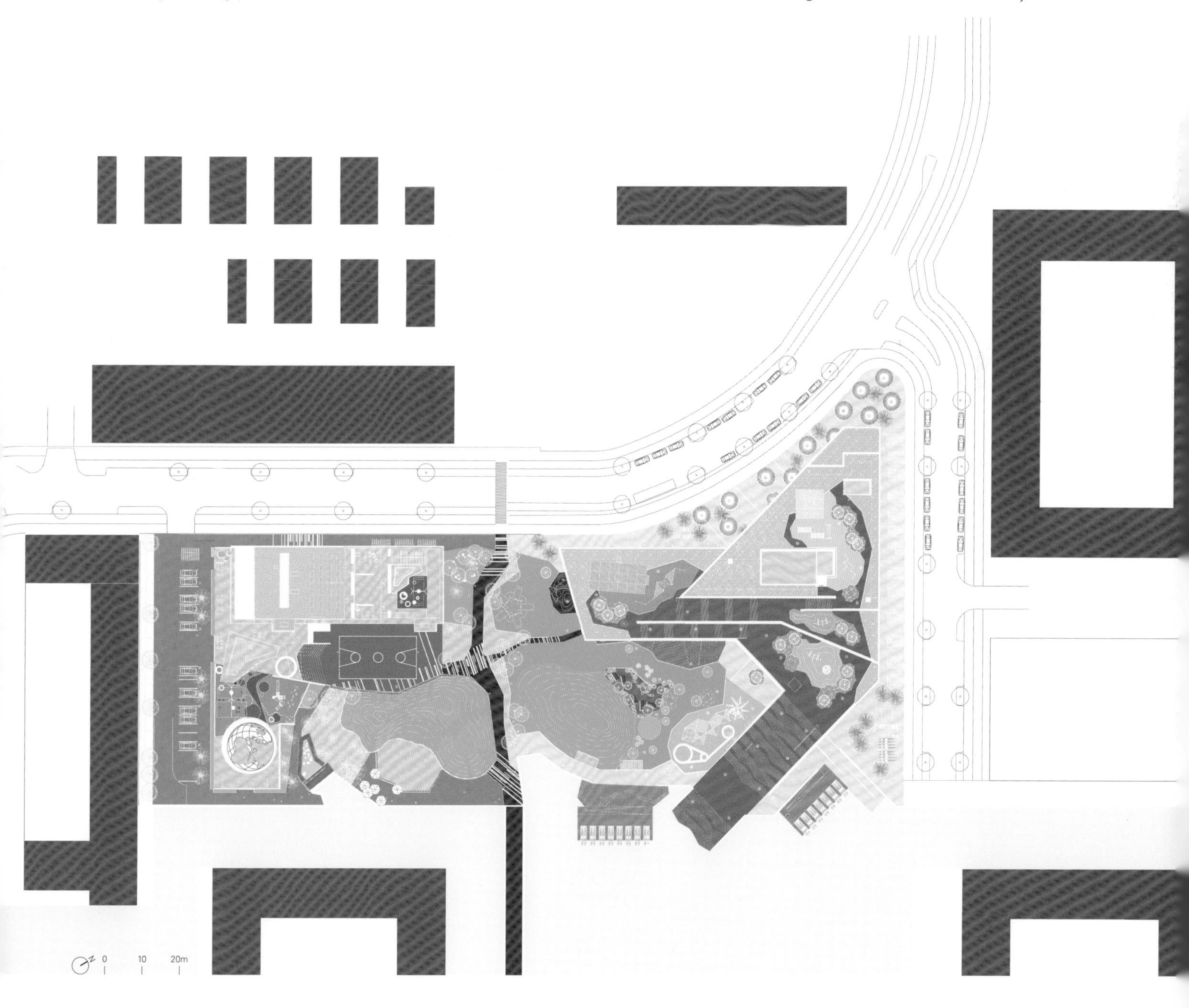

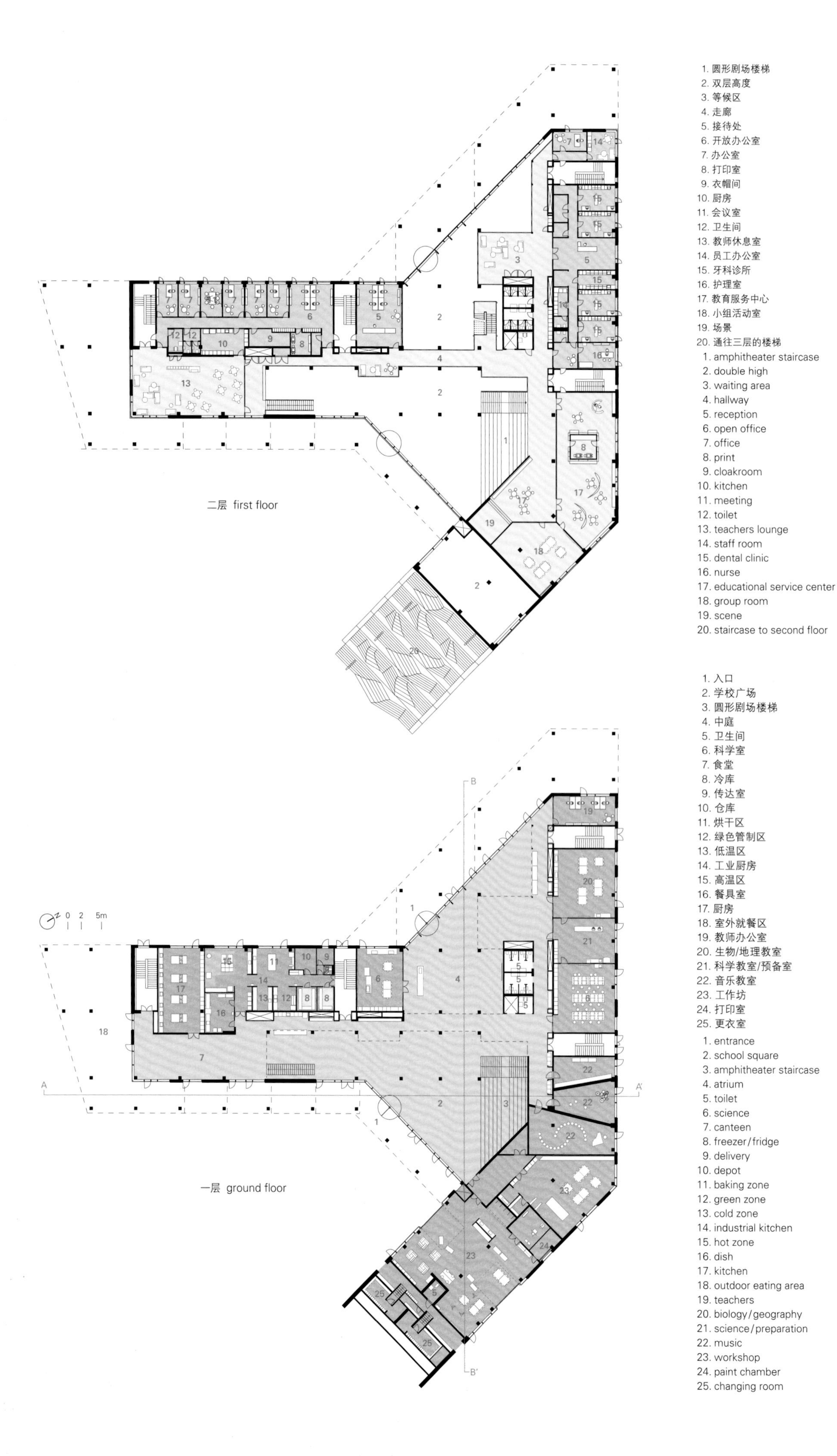

二层 first floor
1. 圆形剧场楼梯
2. 双层高度
3. 等候区
4. 走廊
5. 接待处
6. 开放办公室
7. 办公室
8. 打印室
9. 衣帽间
10. 厨房
11. 会议室
12. 卫生间
13. 教师休息室
14. 员工办公室
15. 牙科诊所
16. 护理室
17. 教育服务中心
18. 小组活动室
19. 场景
20. 通往三层的楼梯
1. amphitheater staircase
2. double high
3. waiting area
4. hallway
5. reception
6. open office
7. office
8. print
9. cloakroom
10. kitchen
11. meeting
12. toilet
13. teachers lounge
14. staff room
15. dental clinic
16. nurse
17. educational service center
18. group room
19. scene
20. staircase to second floor
一层 ground floor
0 2 5m
1. 入口
2. 学校广场
3. 圆形剧场楼梯
4. 中庭
5. 卫生间
6. 科学室
7. 食堂
8. 冷库
9. 传达室
10. 仓库
11. 烘干区
12. 绿色管制区
13. 低温区
14. 工业厨房
15. 高温区
16. 餐具室
17. 厨房
18. 室外就餐区
19. 教师办公室
20. 生物/地理教室
21. 科学教室/预备室
22. 音乐教室
23. 工作坊
24. 打印室
25. 更衣室
1. entrance
2. school square
3. amphitheater staircase
4. atrium
5. toilet
6. science
7. canteen
8. freezer/fridge
9. delivery
10. depot
11. baking zone
12. green zone
13. cold zone
14. industrial kitchen
15. hot zone
16. dish
17. kitchen
18. outdoor eating area
19. teachers
20. biology/geography
21. science/preparation
22. music
23. workshop
24. paint chamber
25. changing room

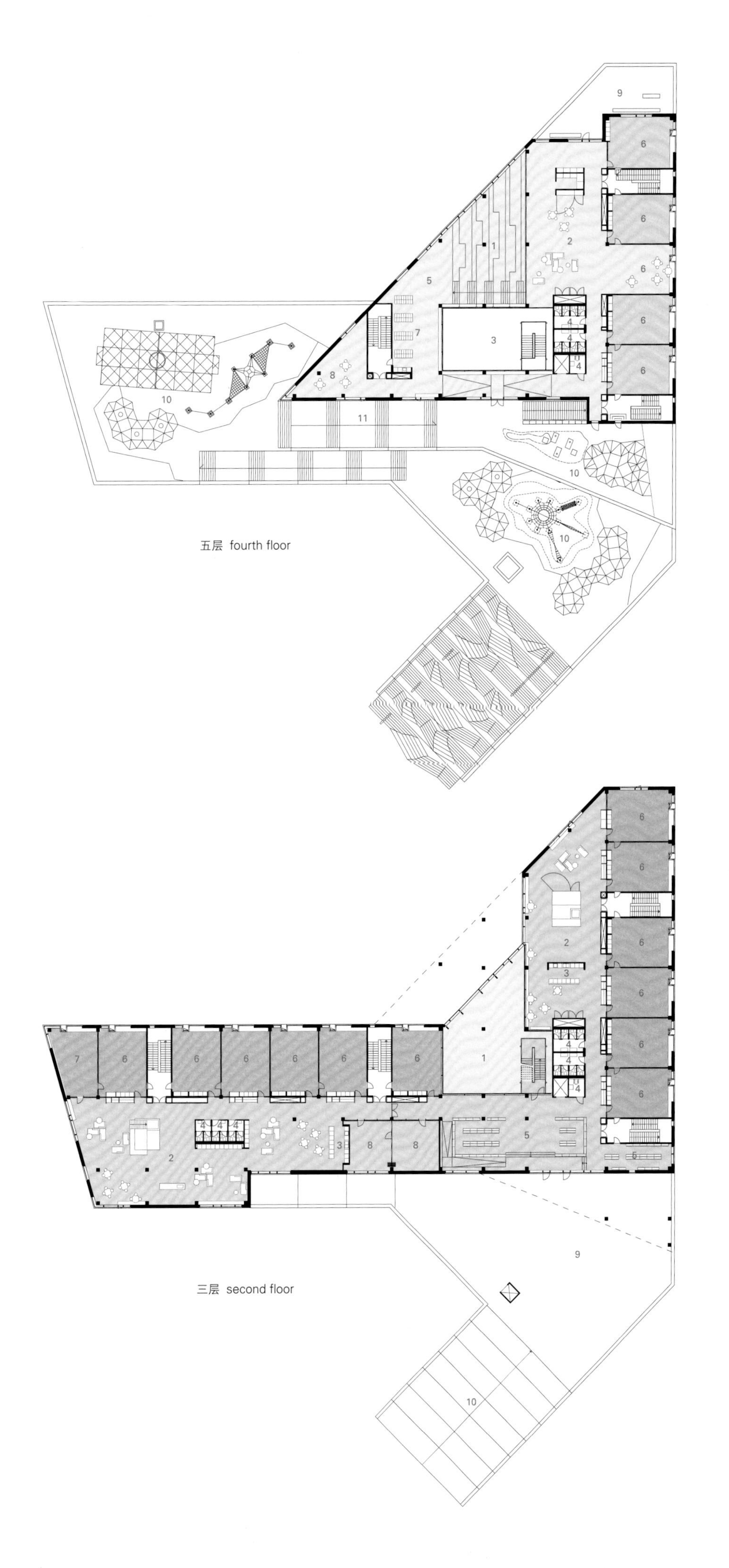

五层 fourth floor

1. 圆形剧场楼梯
2. 公共区域
3. 中庭
4. 卫生间
5. 公共区域的灵活空间
6. 小组活动室
7. 衣帽间
8. 灵活空间
9. 露台
10. 屋顶操场
11. 通往五层的坡道

1. amphitheater staircase
2. common area
3. atrium
4. toilet
5. common area flex zone
6. group room
7. lockers
8. flex zone
9. terrace
10. playground on roof
11. ramp to fourth floor

三层 second floor

1. 中庭
2. 公共区域
3. 厨房
4. 卫生间
5. 衣帽间
6. 教室
7. 会议室
8. 活动室
9. 屋顶操场
10. 通往二层的楼梯

1. atrium
2. common area
3. kitchen
4. toilet
5. wardrobe
6. classroom
7. meeting room
8. activity room
9. playground on roof
10. staircase to first floor

©Stamers Kontor (courtesy of the architect)

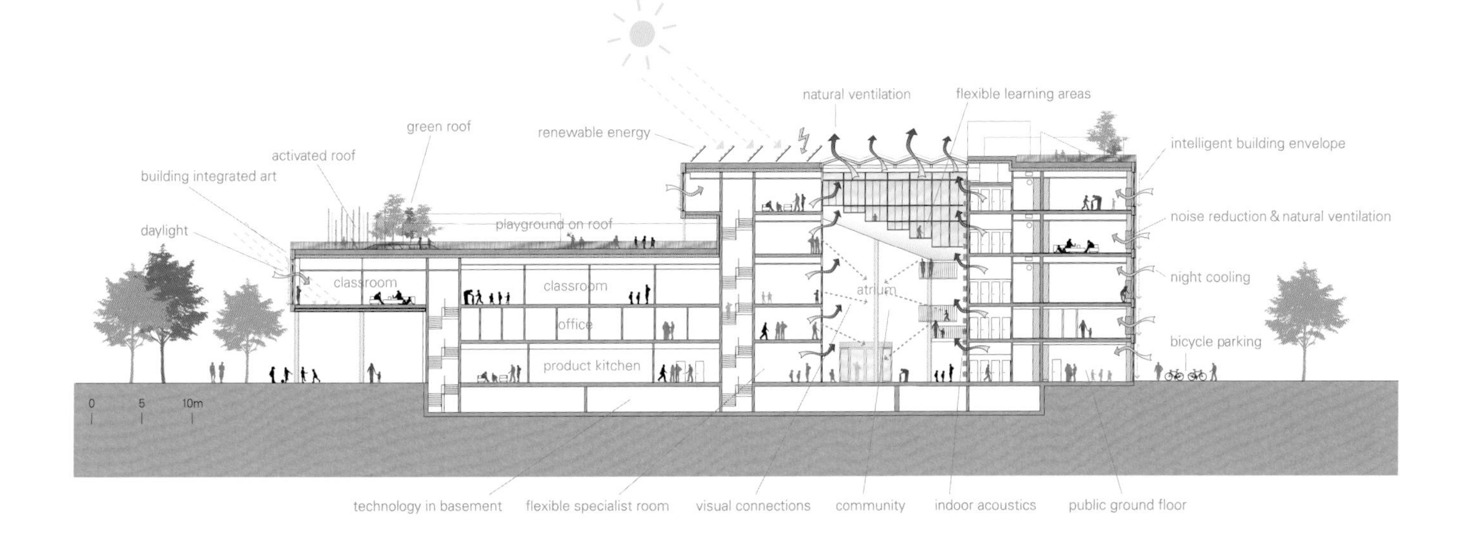

项目名称：South Harbor School / 地点：Støberigade, 2450 København SV, Denmark / 建筑师：JJW Arkitekter / 工程师：NIRAS / 景观设计：JJW Landscape, PK3 Landskab
运动场设计顾问：Keinicke & Overgaard Arkitekter / 总承包商：B. Nygaard Sørensen A/S, G.V.L. ENTREPRISE A/S, Lindpro, Jakon A/S / 施工管理：Friis Andersen Arkitekter
艺术项目：Peter Holst Henckel / 总建筑面积：9,500m² / 竣工时间：2015 / 摄影师：©Torben Eskerod (courtesy of the architect) (except as noted)

schoolyard and city square. A large staircase connects the school with the harbor and is used as an extra classroom, a recreation area for the pupils as well as a vivid public meeting place for the citizens of Copenhagen.

Spaces Adjusted to Pupils Needs

The school is partly divided by age, with the junior classes at the bottom and the older students at the top. The upper floors that house the older pupils are prepared for age-division as well as age-integration. In this way the internal and external spaces take into account what pupils at different stages of their lives are capable of handling and what makes them feel at home and safe. To provide a safe feeling for the younger children, the younger classes have their own home area with small clusters which separate them from the older pupils. The more senior pupils share a large, age-integrated home area spread over two floors where the pupils become part of a larger community across different year-groups.

A Living Facade

The building is characterized by a vivid facade that not only creates variation but also underpins the building's educational function.

The vibrant facade acts as an effective and intelligent building envelope. Automatically openable windows control night cooling, specially developed noise-reducing windows allow natural ventilation, and built facade screens serve both as sun resistance and as an additional heat buffer in the cold season. The architects have taken a relatively common facade material, prefabricated compressed mineral wool paneling, and utilized it for a strong and living expression. The outermost layer consists of 100% recycled aluminum lamellas. The facade changes playfully according to the position of the viewer.

Sustainability

The school follows the parameters of the DGNB approach, which focuses on technical, economic and social sustainability. The use of the building after school hours supports the social aspect of this sustainable approach, for instance the outdoor areas operate as both school yard and public park.

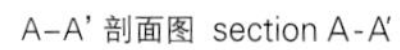
A-A' 剖面图 section A-A'

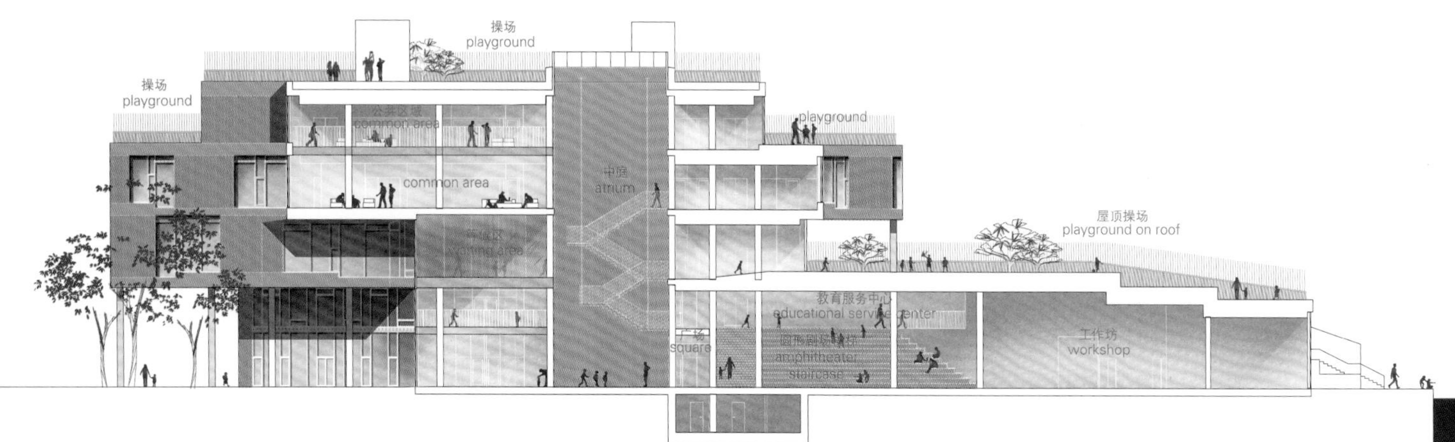

B-B' 剖面图 section B-B'

圣伊西多尔学校坐落在促进可持续发展和生物多样性的尼斯西部的一个生态谷。这些环境因素促使ANMA建筑事务所在设计学校的扩建部分时，谦逊地融入了周围村落的乡村精神。

学校沐浴在法国里维埃拉的阳光下，利用当地的自然资源，学校每天接待数百名儿童走入校园。

社区规模的综合学校

该扩建项目最近竣工，包括6个学龄前教室、包含80个摇篮的托儿中心和一个休闲中心。该场地利用附近街区、钟楼以及周围山丘的地理优势，让人可以直接将美丽的景色尽收眼底。

看第一眼，整座建筑会给人一种谨慎的感觉，但当你通过连续的天井进行深入的体验时，它又因为它的透明性和灯光的设计而变得那么与众不同。因此，它就像一个儿童规模的巨大城市蜂巢。

适合儿童的建筑设计

该项目的体量多样性非常适合于儿童的规模。学校主要有两个楼层：一层整体和二层的一部分主要是功能性建筑，而顶部覆盖着一个轻盈的折叠绿色遮阳屋顶。

植物屋顶和遮阳屋顶的独特结构为整个项目带来了独特的个性，无论是室内还是室外的嬉戏设计，都完美地适应了童年和学校的生活。

圣伊西多尔学校集团扩建工程
Saint Isidore School Group Extension

ANMA

这个蜂巢状的保护设施通过一个由植被构成的边缘设计而与外面的世界区分开来，这些植被恰恰与学生的绘画作品相呼应。整个项目的分布由巨大的玻璃走廊决定，这些走廊将内部生活空间与外界的噪声和炎热隔离开来。

窗户框架和室内空间的明亮颜色定义了三个区域：红色的托儿中心、黄色的休闲中心和橙色的学前区域。

从木质立面到折叠遮阳屋顶

整个屋顶和遮阳屋顶的褶皱为项目提供了一个独特的设计，非常适合儿童和学校。朝北的屋顶上方覆盖着植被，而朝南的屋顶配备有光伏板。

所有的立面都用落叶松木覆盖。正立面全部用金属板覆盖，并配有大门。这些门的顶部覆盖着一条热涂层金属带，这样的设计可以拉长立面。而且，每扇门都按照所在区域的功能用一种颜来表示。

作为项目出发点的自然能源

本项目通过惰性（冷却）混凝土和隔热木材的混合结构采用了可持续设计方法：这样的处理提供了一个非常有效的热舒适系统。此外，从一开始这个设计的目标就是实现零能源消耗：通过限制能源需求和可再生能源的使用来实现这一目标。

地热热泵系统对清洁水进行加热，地下水表可以实现双重目标：在冬季，热量通过地板下供暖在建筑物内传播，而在夏季，循环则通过收集和循环进入建筑物的冷却空气来制冷。屋顶南侧褶皱位置上的光电板提供了电力补充。

由于对自然能源的深入研究，这座低能耗建筑被离散地融入其城市和自然环境中。重新改造的学校现在是一个让周边区域的孩子们学习、成长和探索的地方。

Saint Isidore School sits in an Eco-Valley, located west of Nice, that fosters sustainable development and biodiversity. These environmental considerations led architects ANMA to design the extension to fit humbly into the village spirit of the neighborhood.
Steeped in the sun of the French Riviera, the school draws on local natural resources to welcome hundreds of children every day.

An Integrated School at Neighborhood Scale

Recently completed, this extension is comprised of six preschool rooms, an 80-cradle childcare center and a leisure center. The site takes advantage of immediate views of on the neighborhood and the nearby bell tower as well as surrounding hills.

On first viewing, the architecture of the facility appears discreet, but it becomes extraordinary when experienced through the successive patios, with their transparencies and light effects. Thus, it stands as a gigantic urban hive at a child's scale.

An Architectural Design for Children

The project's volume diversity fits to a child's scale. The school has two main floors: functional volumes on the ground floor and some on the partial first floor, with a light pleated green shade roof sitting on top.
The unique composition of vegetal and shade roofs provides the project with a particular personality, both domestic and playful, perfectly adapted to the world of childhood and school.

La Baronne
Lingostière
Plana de Flori
collet de
Grisella
Plaine du Var
Vallon de
crémat
Saint Isidore
Colle St
Isidore
Les Iscles
Fabron
supérieur
N
0 100 200m

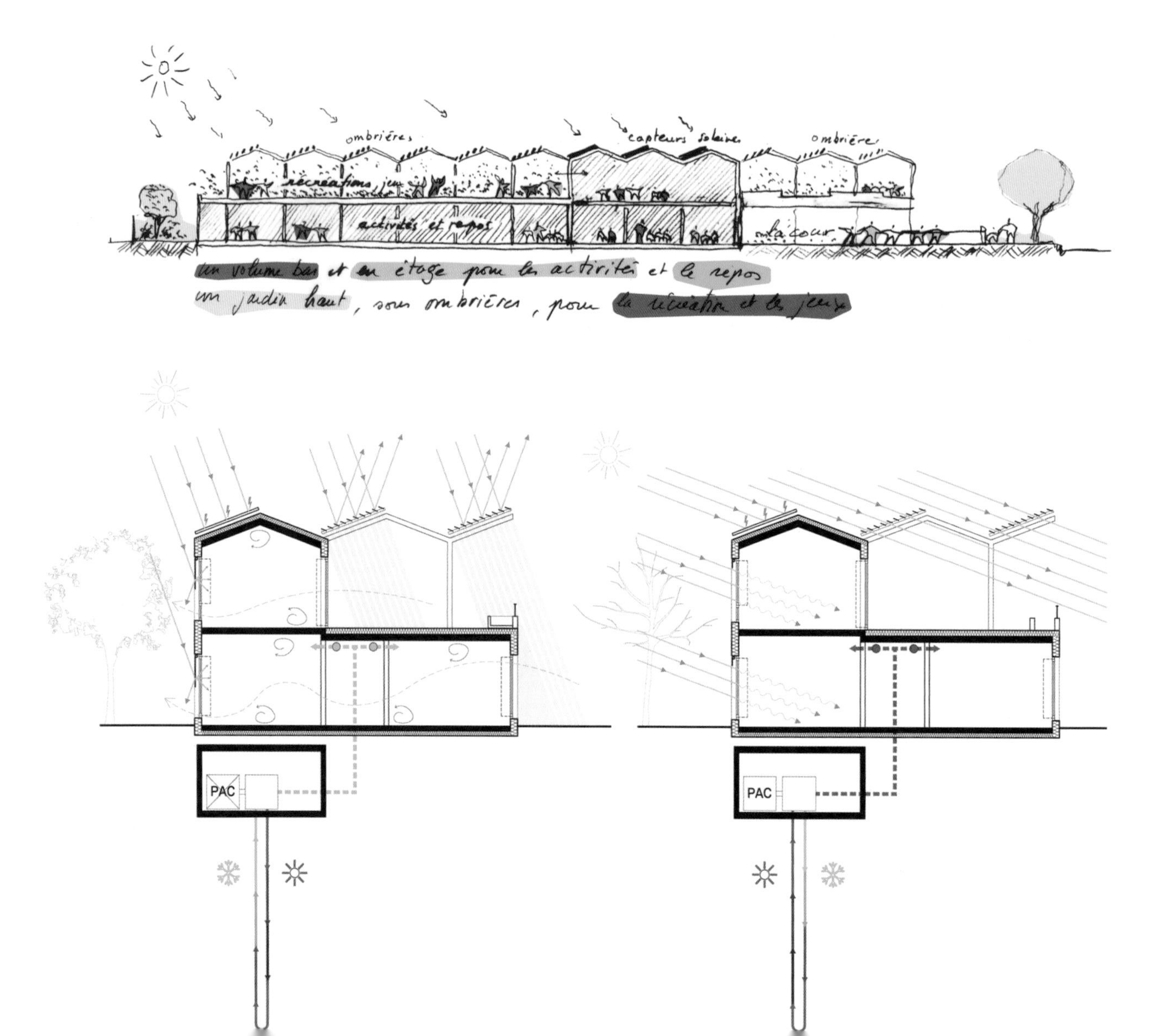
ombrières
capteurs solaires
ombrière
récréations jeu
activités et repos
la cour
un volume bas et un étage pour les activités et le repos
un jardin haut, sous ombrières, pour la récréation et les jeux
PAC
PAC

Liberté
Egalité
Fraternité

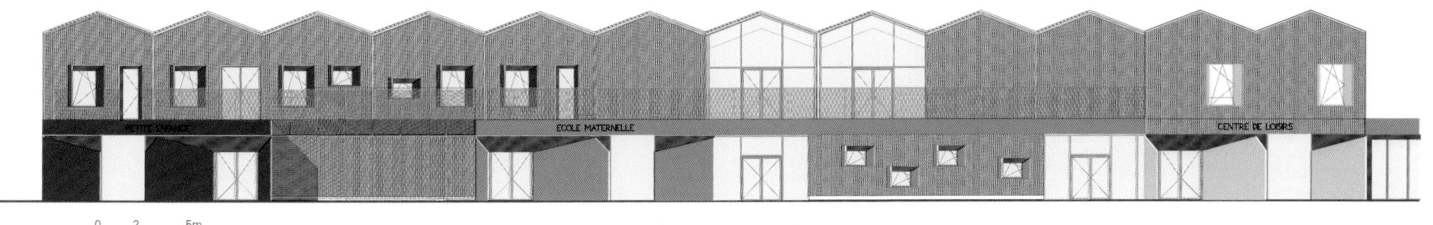

东立面 east elevation

项目名称：Saint Isidore School Group Extension / 地点：18 Avenue Auguste Verola, 06200 NIice, France / 建筑师：ANMA / 项目团队：Valerie Dubois, Antoine Carrel, Hélène Dupont-competition; Helene Galifer, Emmanuel Vinet, Virginie Mira-studies; Alessio Loffredo, Kevin Viel-site / 工程：Batiserf-structural studies; Alto-technical studies; Peutz-acoustic studies / 承包商：SPADA-structure & landscape; CHIRI-facades & metalwork; MSDECO-interior finitions; BUCHET-electric; Eiffage Energie -plumbing, heating, ventilation; Ciel Ascenseurs-elevators; Deal Hydraulique-fountains / 客户：Metropole Nice Cote d'Azur / 总建筑面积：4,500m² / 竣工时间：2017 / 摄影师：©Vincent Fillon (courtesy of the architect)

This hive-like protective facility is demarcated from the outside world by a fringe of vegetation that responds to class drawings. The overall program distribution is determined by great glass corridors that isolate the inner living spaces from sound and heat.
Bright colors on the windows' frames and interiors spaces define the three areas: a red childcare center, a yellow leisure center and an orange preschool.

From Wooden Facades to Pleated Shade Roof

The overall roofing and the shade roof pleats provide the project with a unique design that is adapted to childhood and school. The north facing roofs are covered in vegetation, and south facing roofs are equipped with photovoltaic panels.

All the facades are clad in larch wood. Front facades are entirely covered with sheet metal and equipped with a large portal. The upper part of these portals is topped with a thermal-coated sheet metal band that elongates the facades. Once again, each portal is outlined by a color according to its program.

Natural Energy as Project Engine

The project uses sustainable design methods through a mixed structure of concrete for inertia (coolness) and wood for thermic isolation: together, they provide a very effective system in terms of thermal comfort. Furthermore, a zero energy goal was the aim from the outset: this was achieved through the capping of energy needs and the use of renewable energies.

1. 洗手间（接待处）
2. 健身室（接待处）
3. 儿童图书馆
4. 活动室（托儿所）
5. 多功能室（托儿所）
6. 大厅
7. 校长办公室
8. 教室办公室
9. 活动室、工作坊
10. 宿舍
11. 健身室（幼儿园）
12. 洗手间（幼儿园）
13. 游戏室
14. 活动室（接待处）
15. 活动室（一年级）
16. 洗手间（一年级）
17. 多功能室（一年级）
18. 活动室、工作坊

1. rest room (reception)
2. exercise room (reception)
3. child library
4. activity room (creche)
5. multipurpose room (creche)
6. lobby
7. director's office
8. teacher's office
9. activity workshop room
10. dormitory
11. exercise room (nursery)
12. rest room (nursery)
13. game room
14. activity room (reception)
15. activity room (year 1)
16. rest room (year 1)
17. multipurpose room (year 1)
18. activity workshop room

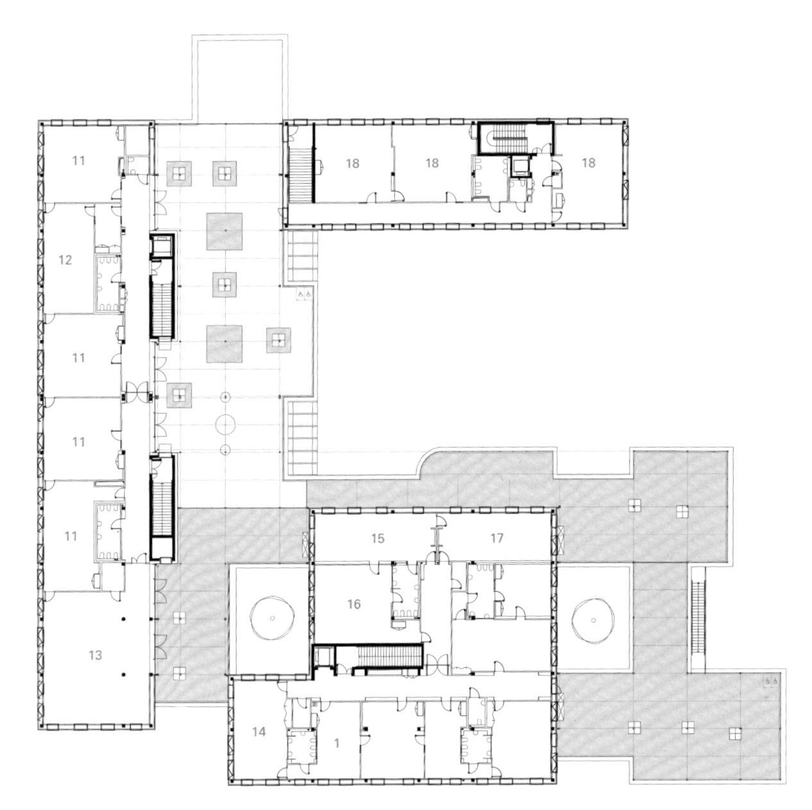

二层 first floor

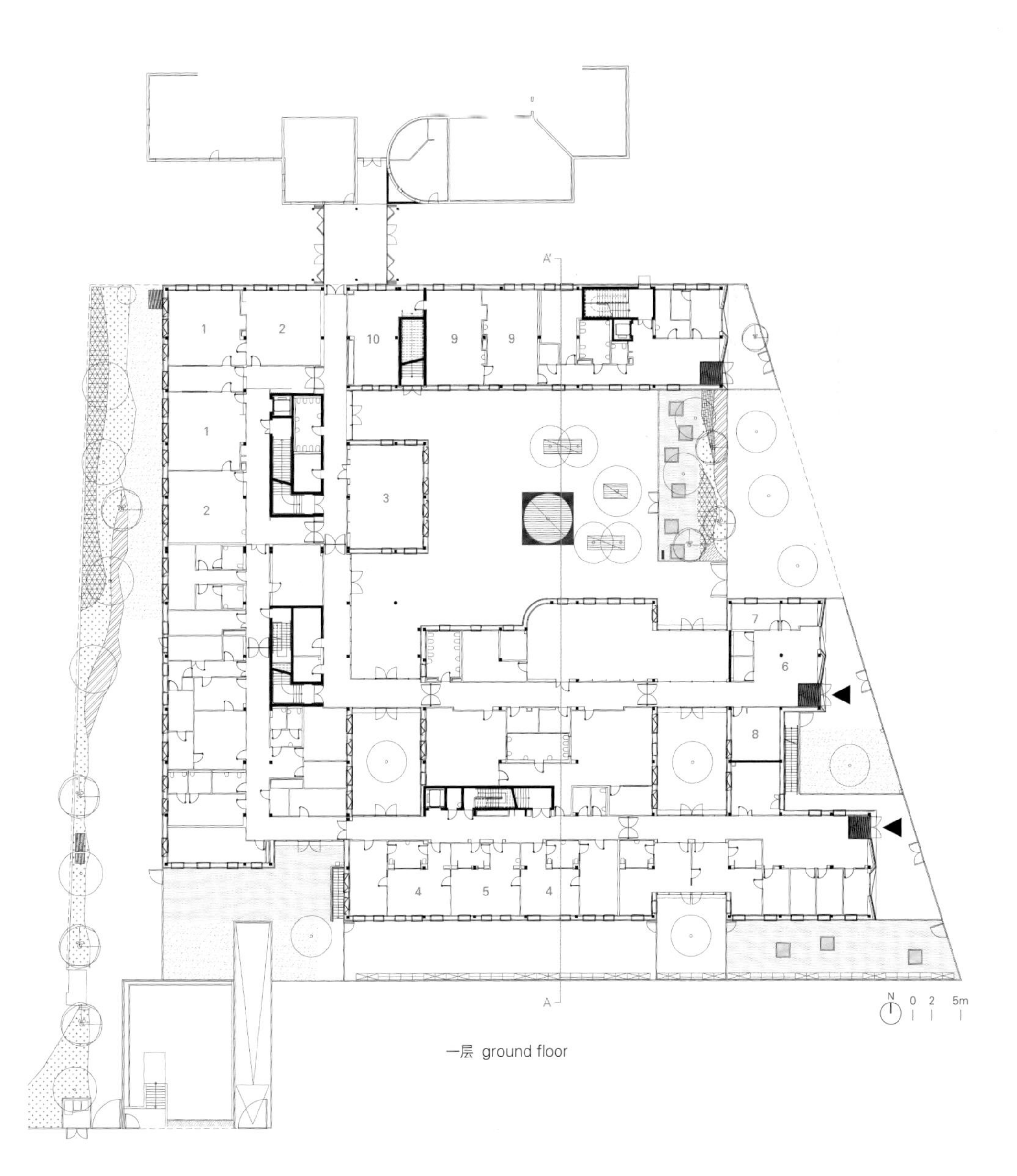

一层 ground floor

A geothermal heat pump system heats the sanitary water, and the phreatic table achieves a double goal: during winter, heat spreads in the building through underfloor heating while, during summer, the cycle goes the other way by gathering and circulating cool air into the building. Photovoltaic panels on the southern pleats of the roof provide complementary electricity production.

Resulting from in-depth studies on natural energies, this low energy concumption building is discretely integrated into its urban and natural context. The reinvented school is now a place for the children of the neighborhood to learn, grow and discover.

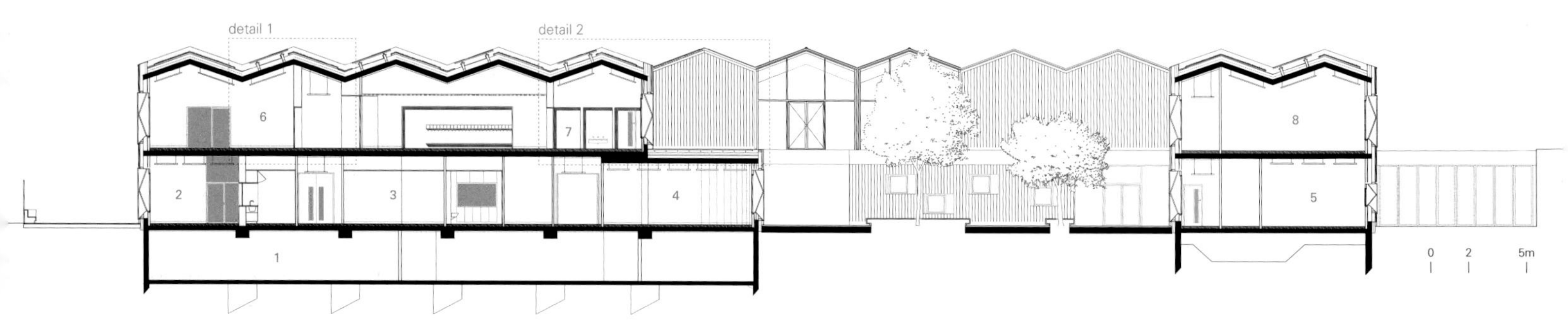

1. 储藏室（托儿所） 2. 活动室（托儿所） 3. 入门教室（托儿所&小学） 4. 游戏室 5. 会议室 6. 洗手间 7. 多功能室 8. 活动室、工作坊
1. storage room (creche) 2. activity room (creche) 3. gateway class (creche & primary) 4. game room 5. meeting room 6. rest room 7. multipurpose room 8. activity workshop room
A-A' 剖面图 section A-A'

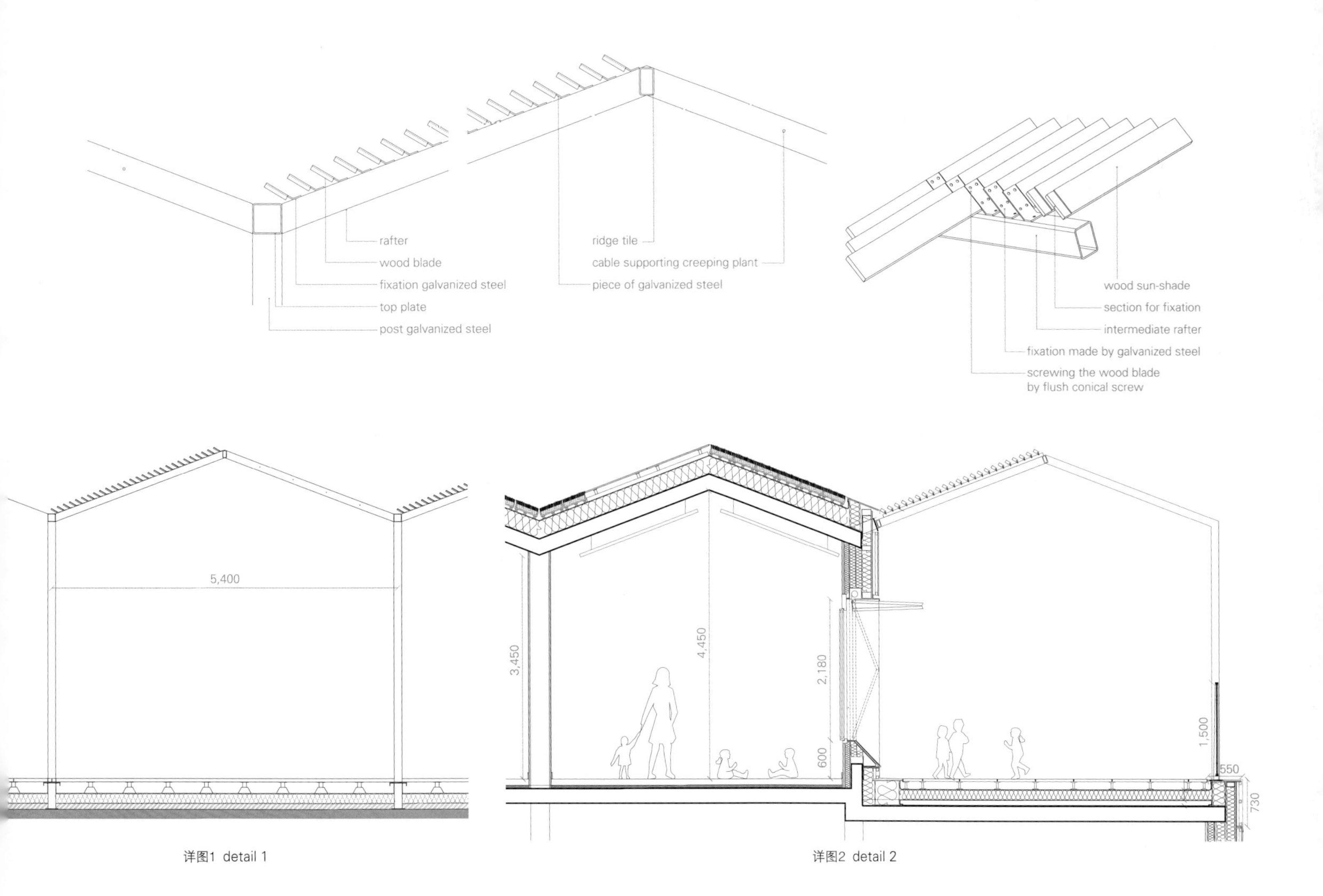

详图1 detail 1

详图2 detail 2

Znjan-Pazdigrad 小学

Žnjan – Pazdigrad Primary School

x3m

Žnjan-Pazdigrad小学项目位于斯普利特郊区，20年来，这一地区一直处于由农业区向“宜居”居民区过渡的过程当中。由于附近缺乏公共设施和空间，学校不仅被设计成教育场所，而且还被设计用于整个社区的公共空间。

该项目包括为720名学生所实施的广泛的教育方案：8个教室（一至四年级），12个学科教学教室（五至八年级），此外还包括有舞台、图书馆、餐厅、厨房，以及一个大型和一个小型的体育馆，它们主要用于体育课、室外操场和运动场。

课堂教学空间、学科教学空间，以及一个大型的体育馆，构成了建筑的三个主要体量，而其他空间和功能的设计则介于它们之间，并在四个楼层上相互联系。垂直的组织体系有利于功能的实施：地下室作为体育运动的空间，一层作为中央集合空间，并有公共通道将学校与体育馆的入口分开，第二层和第三层留作教学使用。

在周边环境中，住宅小区、家庭住宅、小作坊、购物中心、花园、温室与学校和平共处。但是整个地区的公共空间和社会设施极度匮乏。这所新建的小学是一个让人期待已久的项目：当地居民抗议拖延，要求立即动工建设，这次抗议活动也促进了对当地居民的社区意识的塑造。该场地位于陡峭的下坡地形上，区域太小，不利于所需项目的建设，因此为了尽量增加可利用空间，建筑师创建了一个新的地形。建筑师使用了大量的平台互相叠加，使学校区域和体育设施与开放的多功能空间相互交织，而整个空间结构也可以激发使用者对室内外活动的体验。

在整个设计过程中，获得的公共空间面积（12100m²）超过了整个场地面积（11600m²），为社区提供了大量的室外公共活动空间。

本项目的设计基于暴露的结构：柱、板、百叶窗和墙体在结构上支撑着建筑，同时作为其立面和最重要的内部元素。学校的整体外观

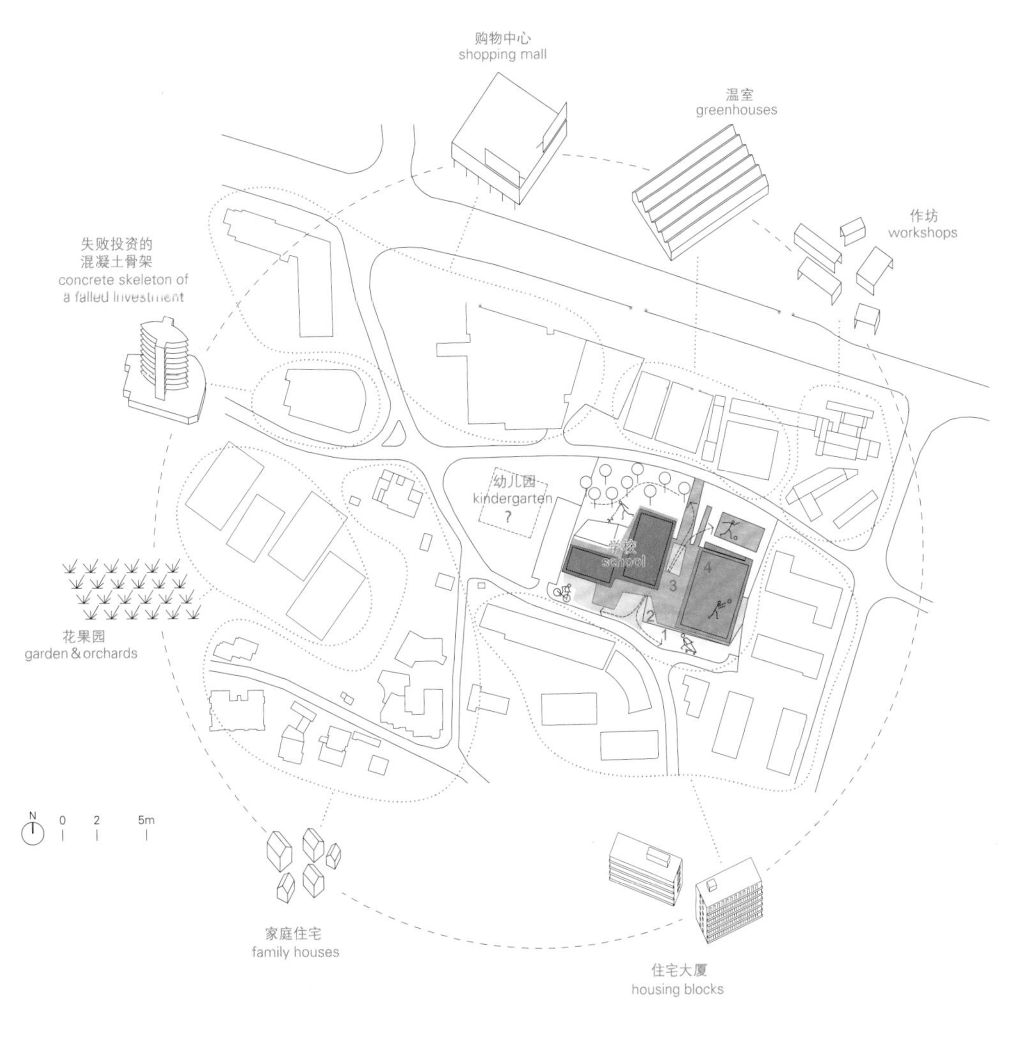

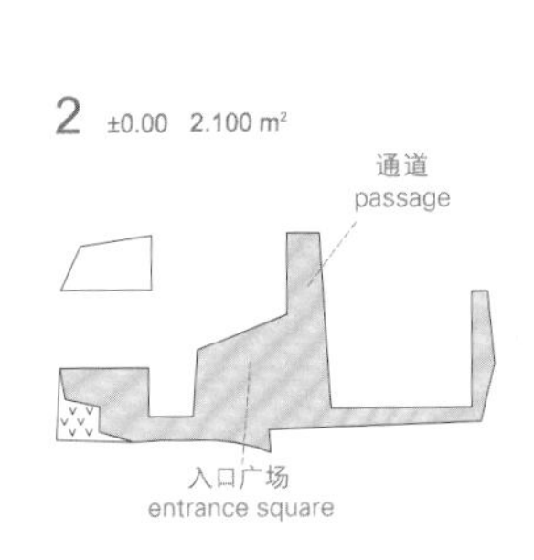

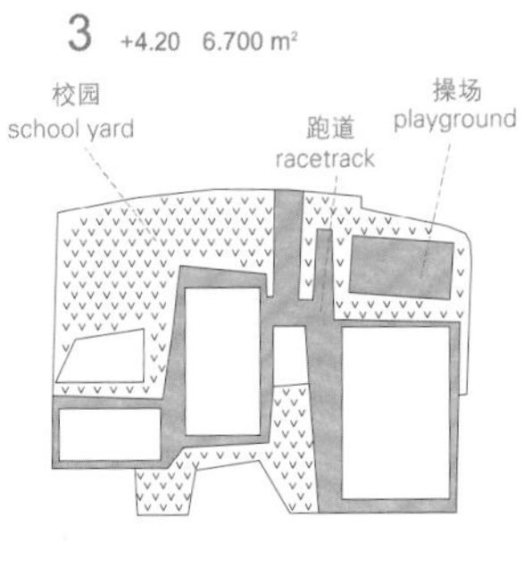

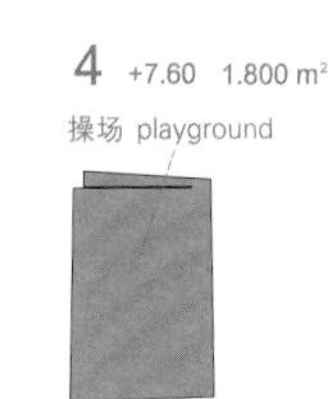

为浅灰色浇注混凝土；内部的地下室和一层的交流区域、大厅都铺设了抛光水磨石地面。教室和教室走廊使用细木橡木地板，与家具和隔声天花板的设计相得益彰。室内和室外体育馆的木地板，再加上室内的装饰，使室内环境显得更加中性。

Nikola Đurek对立面的字体设计提供了一种额外的含义，为这个“有意保留未完工状态”的工程起到了保护“表皮”的作用。

The Žnjan - Pazdigrad Primary School project is located on the outskirts of Split, in an area that has for at the least 20 years been in transition from agricultural suburbia to a "proper" residential quarter. Due to the lack of public amenities and spaces in the near vicinity, the school is designed not just as a place of education, but also as a collective space for the entire neighborhood.

The project comprises an extensive educational program for 720 pupils: 8 teaching classrooms (for first to fourth grade), 12 subject teaching classrooms (for fifth to eighth grade),

an assembly hall with a stage, a library, a dining area and kitchen, a large and a smaller sports hall for physical education, outdoor playgrounds and an athletics field.
The learning spaces for class teaching and subject teaching, and a large sports hall, constitute three volumes, while the other spaces and programs are fluidly organized in between, and are interconnected on four levels. The vertical organization facilitates the functional one: the basement serves as a space for gymnastics and sports, the ground floor as a central assembly space with a public passage separating the school from the sports hall entrance, and the first and second floors are reserved for teaching and studying.
In the peripheral surroundings, housing blocks, family houses, small workshops, shopping malls, gardens and greenhouses peacefully coexist, but the entire area suffers

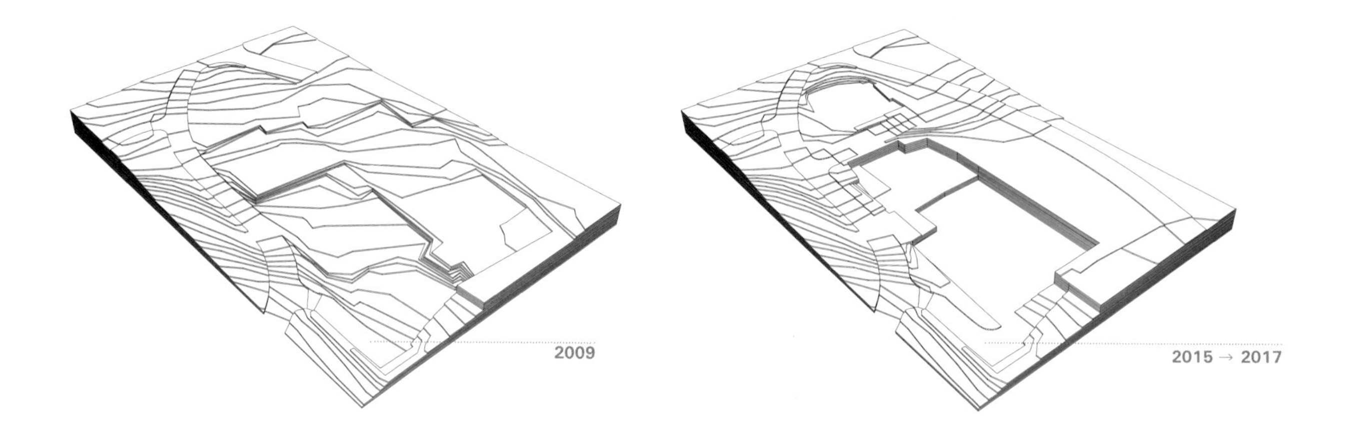
2009
2015 → 2017

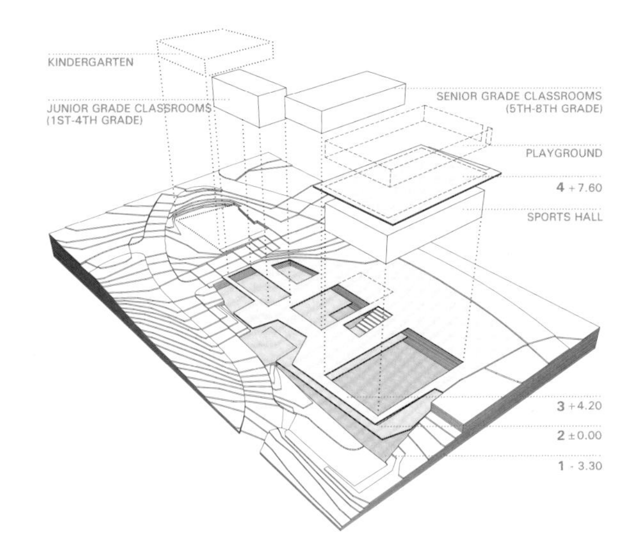
KINDERGARTEN
JUNIOR GRADE CLASSROOMS (1ST-4TH GRADE)
SENIOR GRADE CLASSROOMS (5TH-8TH GRADE)
PLAYGROUND
4 +7.60
SPORTS HALL
3 +4.20
2 ±0.00
1 - 3.30

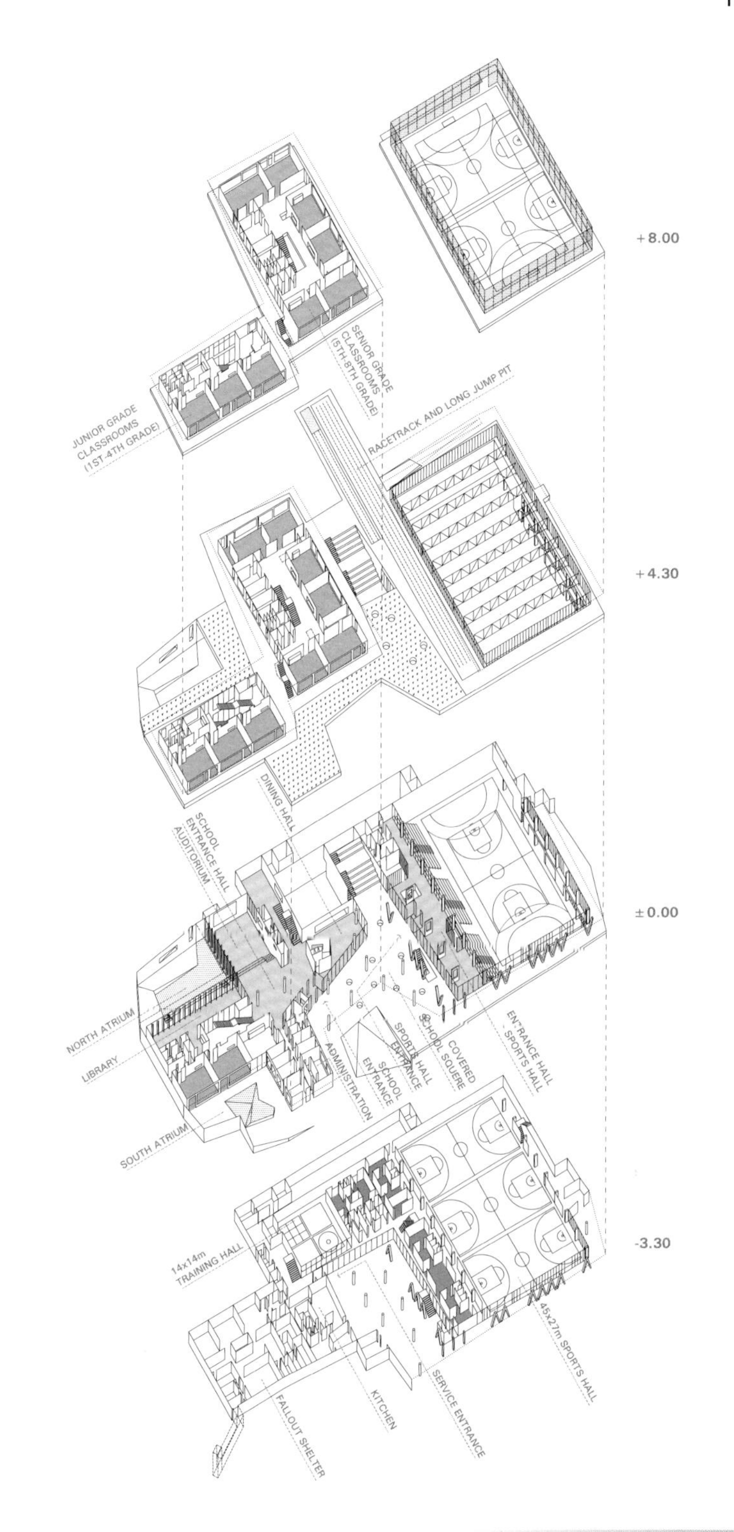
+8.00
SENIOR GRADE CLASSROOMS (5TH-8TH GRADE)
JUNIOR GRADE CLASSROOMS (1ST-4TH GRADE)
RACETRACK AND LONG JUMP PIT
+4.30
DINING HALL
SCHOOL ENTRANCE HALL
AUDITORIUM
±0.00
NORTH ATRIUM
LIBRARY
SOUTH ATRIUM
ADMINISTRATION
SCHOOL ENTRANCE
SPORTS HALL ENTRANCE
COVERED SCHOOL SQUARE
ENTRANCE HALL SPORTS HALL
-3.30
14x14m TRAINING HALL
FALLOUT SHELTER
KITCHEN
SERVICE ENTRANCE
45x27m SPORTS HALL

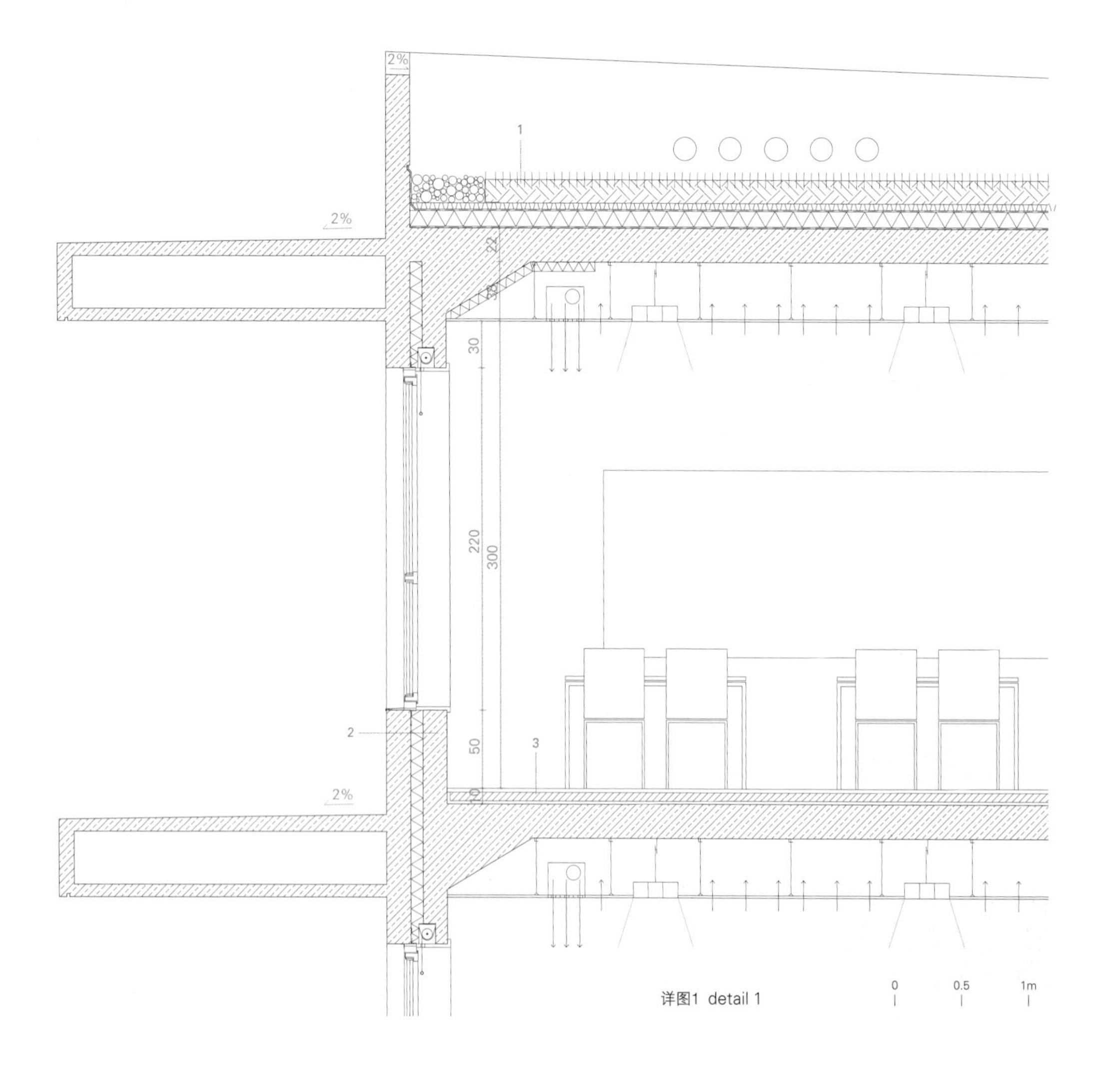

详图1 detail 1

1. 15cm extensive green roof
   0.2cm filter mat
   5.0cm drainage layer
   0.2cm filter mat
   0.2 + 0.2cm waterproofing TPO membrane
   10cm extruded polystyrene (XPS) thermal insulation
   0.5cm vapour barrier
   22cm reinforced concrete
   37cm ceiling suspension (lighting, heating, cooling, fire sprinklers)
   1.25cm perforated acoustic plasterboard
2. 16cm reinforced concrete, water-resistant
   8cm expanded polystyrene (EPS) thermal insulation
   16cm reinforced concrete
3. 2.5cm solid oak parquet
   5.5cm screed
   0.02cm polyethylene sealing layer
   2x2cm expanded polyethlene (EPS) board
   22cm reinforced concrete
   37cm ceiling suspension (lighting, heating, cooling, fire sprinklers)
   1.25cm perforated acoustic plasterboard

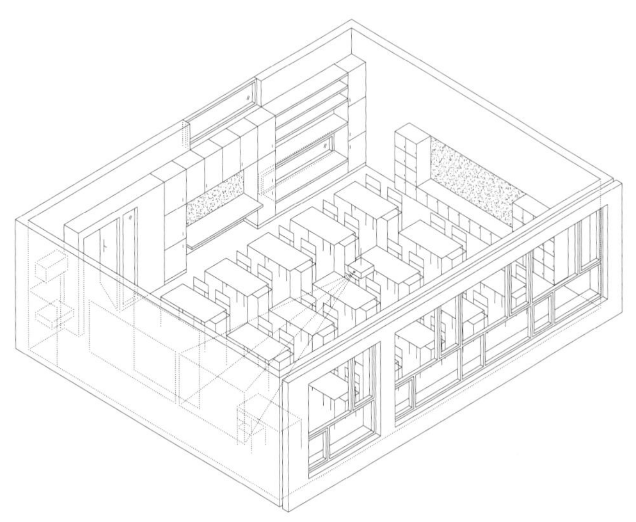
低年级教室（一到四年级）
junior grade classroom (1st~4th grade)

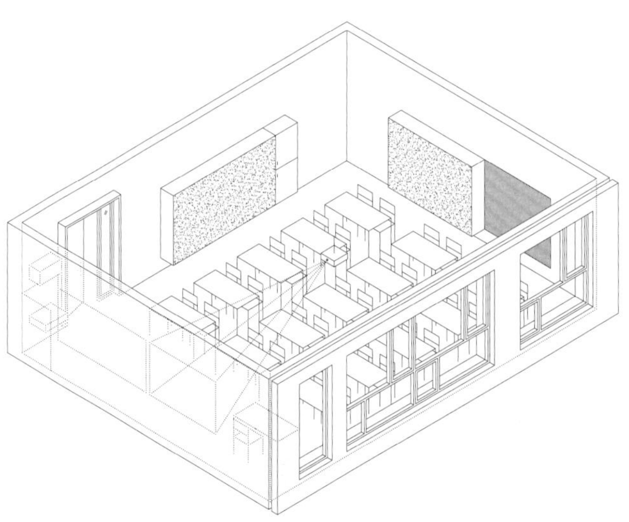
高年级教室（五到八年级）
senior grade classroom (5th~8th grade)

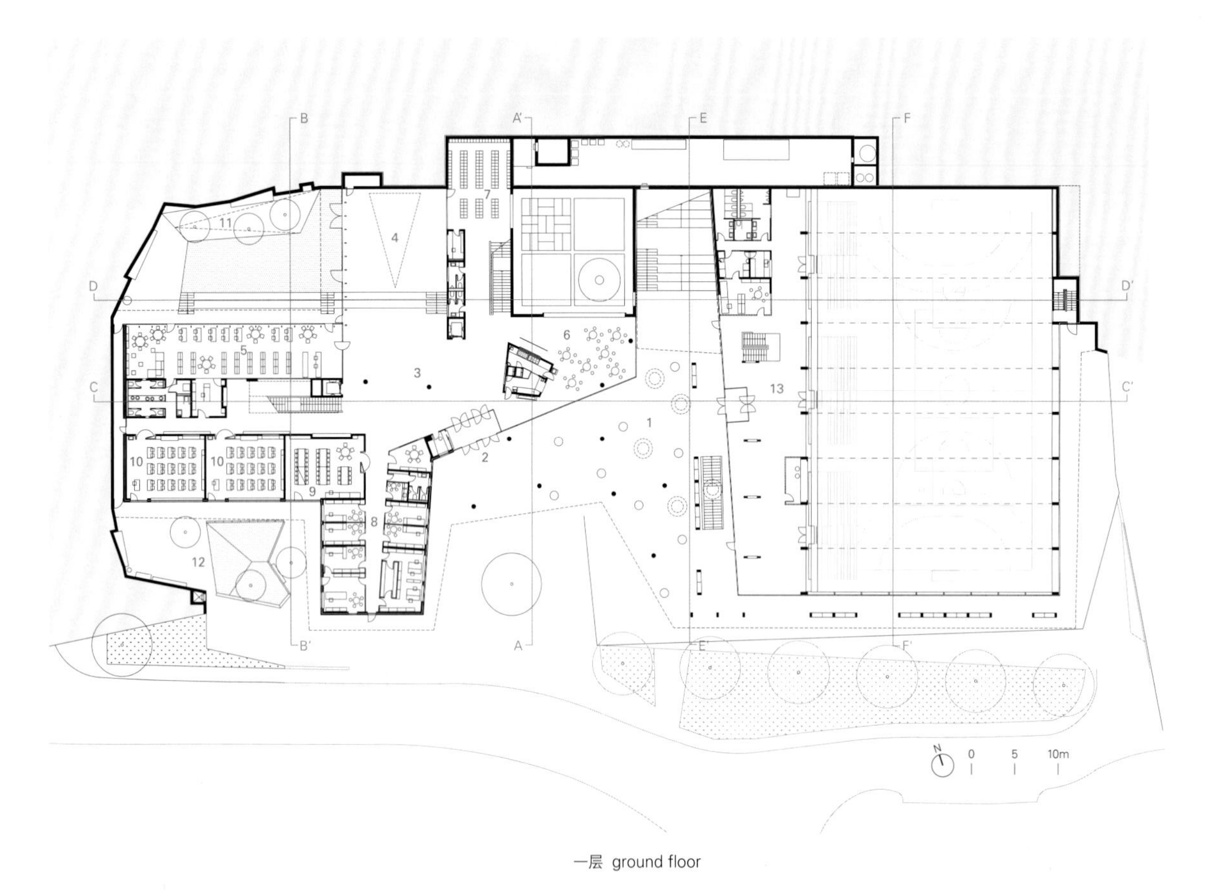

一层 ground floor

1. 带顶的学校前院
2. 学校入口
3. 学校入口大厅
4. 礼堂
5. 图书馆
6. 餐厅
7. 更衣室
8. 学校管理办公室
9. 员工办公室
10. 低年级教室
11. 北部庭院操场
12. 南部庭院操场
13. 入口大厅–体育馆

1. covered school forecourt
2. school entrance
3. school entrance hall
4. assembly hall
5. library
6. dining hall
7. locker
8. school administration
9. staffroom
10. junior grade classroom
11. north courtyard playground
12. south courtyard playground
13. entrance hall–sports hall

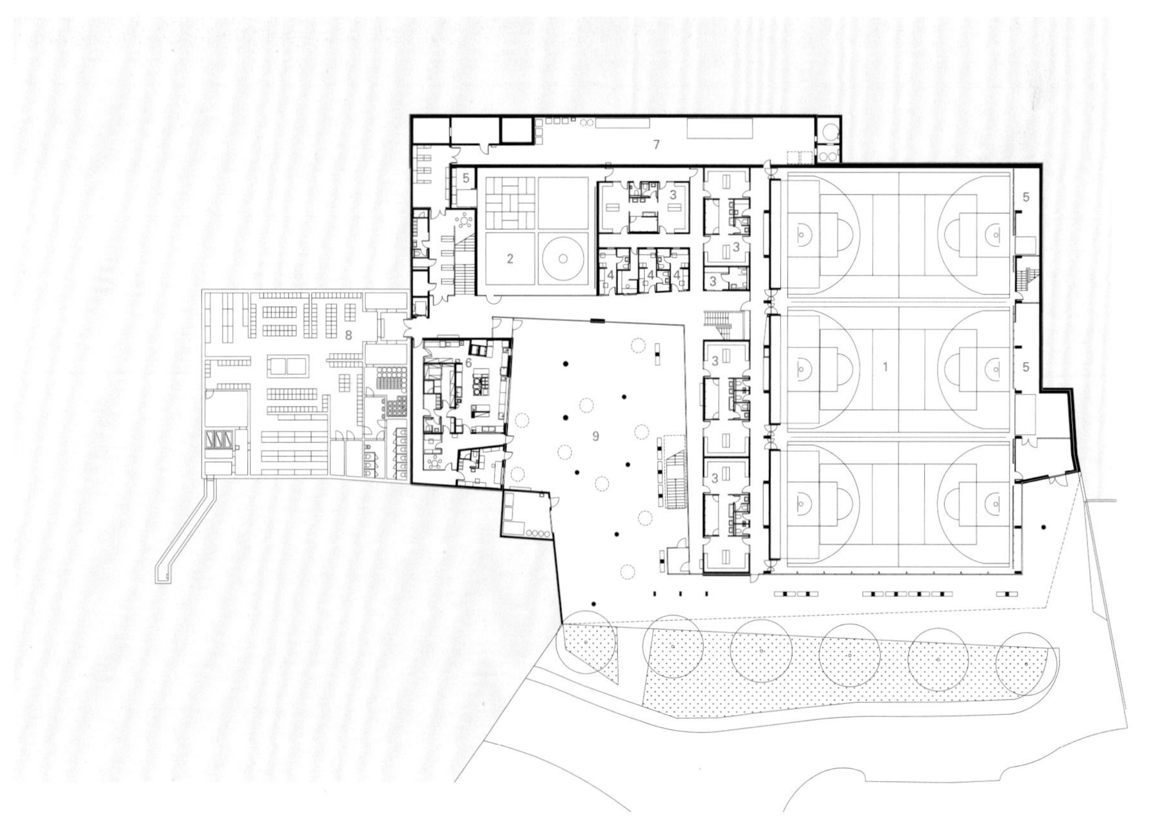

地下一层 first floor below ground

1. 体育馆（45m×27m）
2. 训练馆（14m×14m）
3. 更衣室
4. 教师更衣室
5. 体育器材室
6. 厨房
7. 机械设备室
8. 微粒掩蔽所
9. 设备入口

1. sports hall (45x27m)
2. training hall (14x14m)
3. changing room
4. teacher's changing room
5. sports equipment storage
6. kitchen
7. mechanical room
8. fallout shelter
9. service entrance

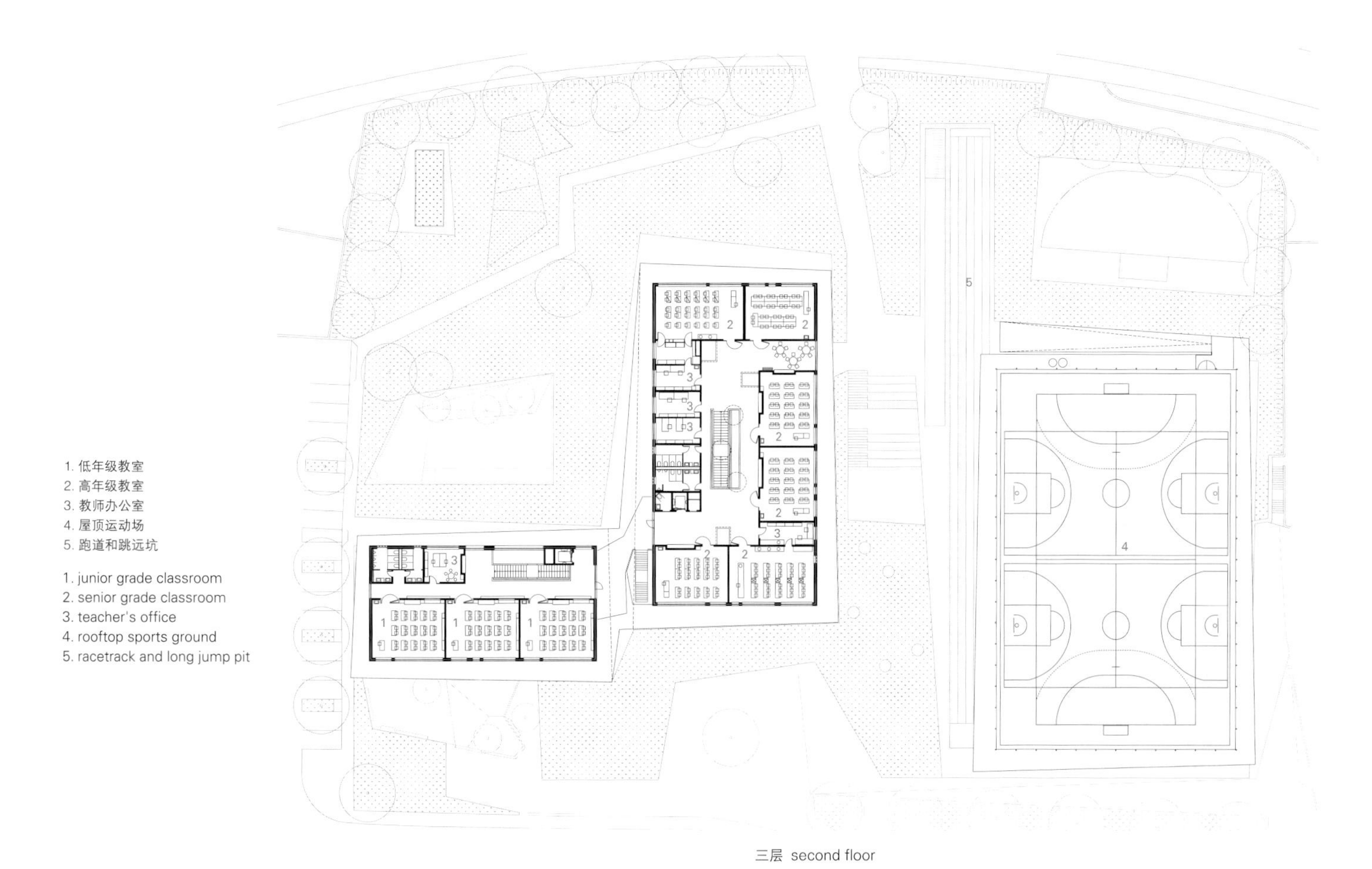

三层 second floor

南立面 south elevation

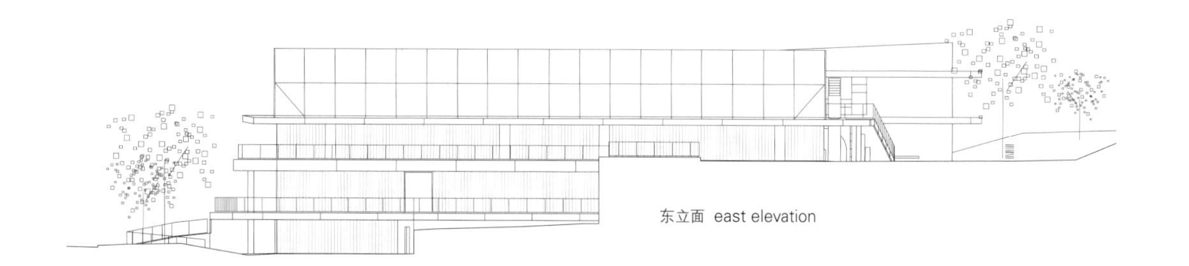

东立面 east elevation

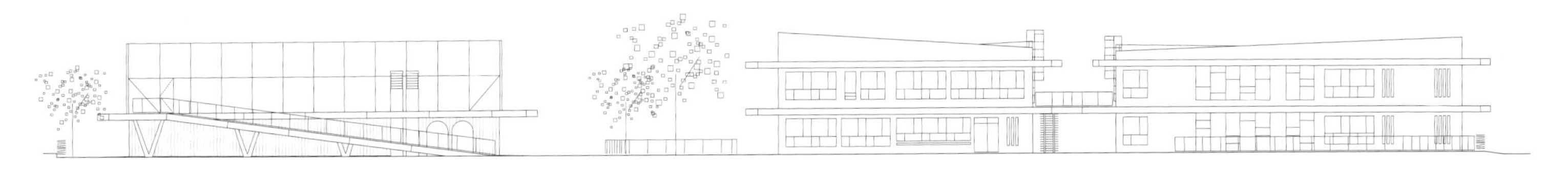

北立面 north elevation

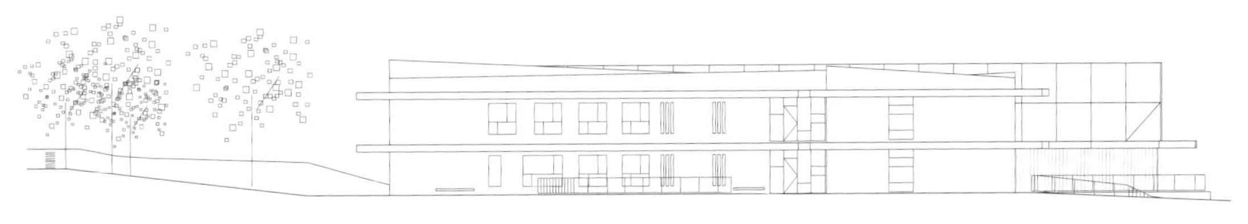

西立面 west elevation

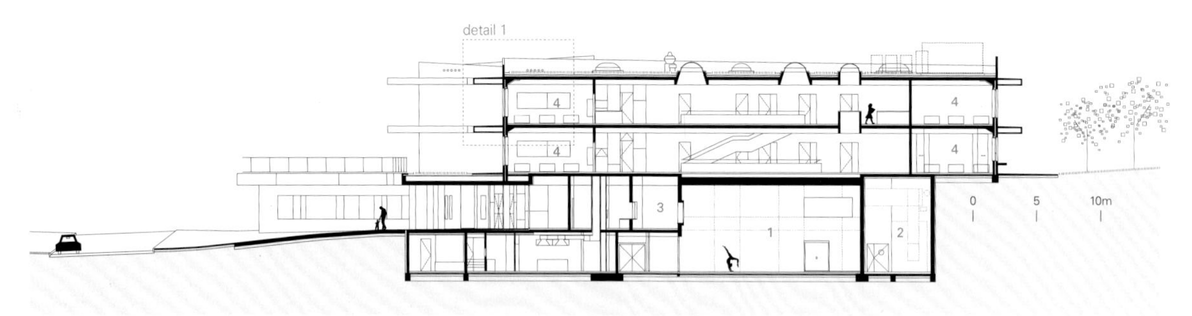

1. 训练馆 2. 机械设备室 3. 餐厅 4. 高年级教室
1. training hall 2. mechanical room 3. dining hall 4. senior grade classroom
A-A' 剖面图 section A-A'

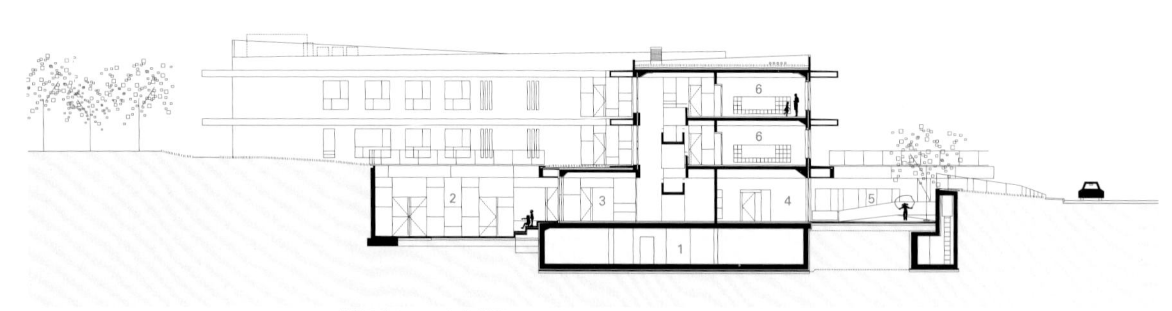

1. 微粒掩蔽所 2. 北部庭院操场 3. 图书馆 4. 员工办公室 5. 南部庭院操场 6. 低年级教室
1. fallout shelter 2. north courtyard playground 3. library 4. staff room 5. south courtyard playground 6. junior grade classroom
B-B' 剖面图 section B-B'

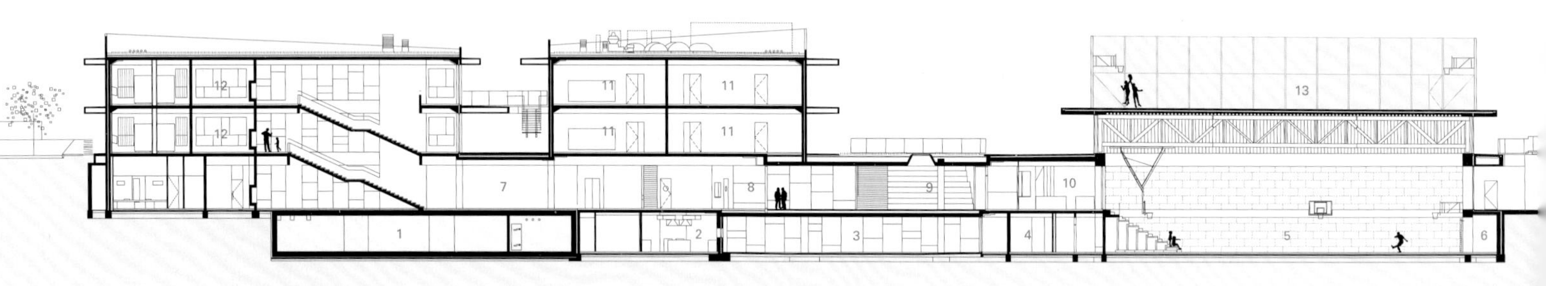

1. 微粒掩蔽所 2. 厨房 3. 设备入口 4. 更衣室 5. 体育馆 6. 体育器材室 7. 学校入口大厅 8. 餐厅 9. 带顶学校前院
10. 入口大厅-体育馆 11. 高年级教室 12. 教师办公室 13. 屋顶运动场
1. fallout shelter 2. kitchen 3. service entrance 4. changing room 5. sports hall 6. sports equipment storage 7. school entrance hall
8. dining hall 9. covered school forecourt 10. entrance hall-sports hall 11. senior grade classroom 12. teacher's office 13. rooftop sports ground
C-C' 剖面图 section C-C'

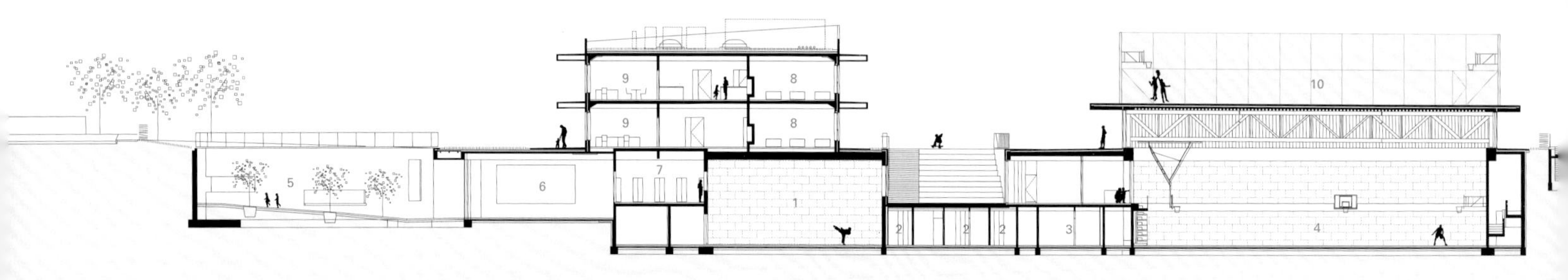

1. 训练馆 2. 教师更衣室 3. 更衣室 4. 体育馆 5. 北部庭院操场 6. 礼堂 7. 更衣室 8. 高年级教室 9. 教师办公室 10. 屋顶运动场
1. training hall 2. teacher's changing room 3. changing room 4. sports hall 5. north courtyard playground
6. assembly hall 7. locker 8. senior grade classroom 9. teacher's office 10. rooftop sports ground
D-D' 剖面图 section D-D'

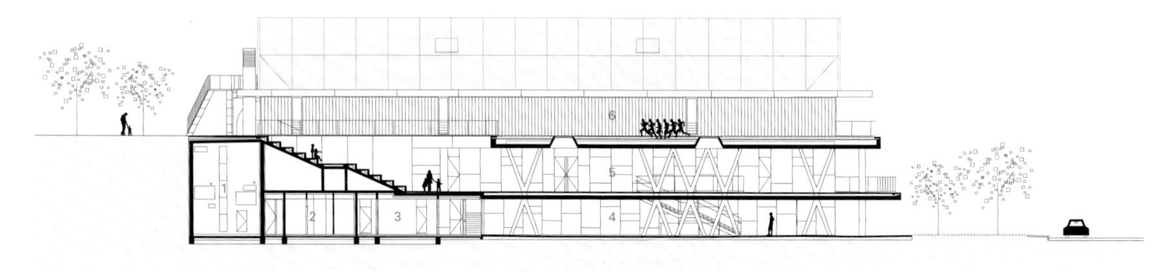

1. 机械设备室 2. 更衣室 3. 教师更衣室 4. 设备入口 5. 带顶学校前院 6. 跑道和跳远坑
1. mechanical room 2. changing room 3. teacher's changing room 4. service entrance 5. covered school forecourt 6. racetrack and long jump pit
E-E' 剖面图 section E-E'

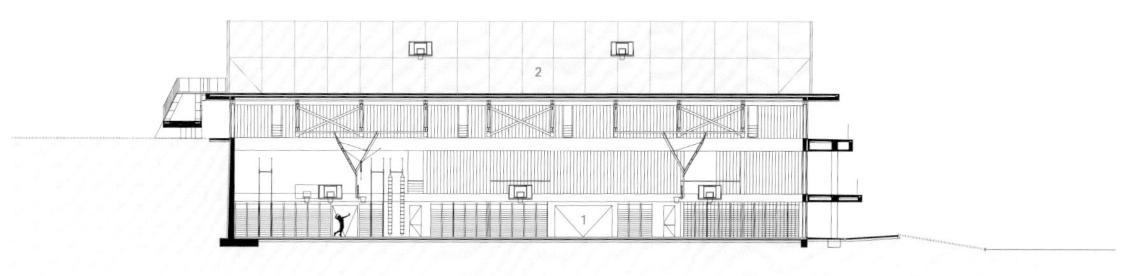

1. 体育馆 2. 屋顶运动场
1. sports hall 2. rooftop sports ground
F-F' 剖面图 section F-F'

from an extreme shortage of public spaces and social amenities. The new primary school was a long-awaited project: protests demanding its immediate construction helped to shape a sense of community among the local residents. The site, located on a steep and descending terrain, was too small to facilitate the required program, so a new topography was created to maximize and multiply the available space. A series of platforms, or decks, one above another, enable the school areas and sports facilities to intertwine with open multipurpose spaces, while the whole spatial structure serves as a generator of indoor and outdoor activities and experiences.
Throughout the design process, the public space area obtained (12,100m²) has become larger than the entire site area (11,600m²), offering the community a wide array of outdoor collective spaces.

The project design is based on exposing the structure: columns, slabs, brise-soleils and walls support the building structurally, simultaneously acting as its facade and most important interior element. Light grey cast concrete dominates the overall appearance; the interior is complemented with polished terrazzo in the basement and ground floor communication areas and halls, oak wood flooring in the classrooms and classroom corridors, joinery, integrated furniture and acoustic ceilings. Color accents in an otherwise rather neutral environment are to be found in the indoor and outdoor sports hall floorings, coupled with the interior finishes.
Nikola Đurek's typographical intervention on the facade provides an additional semantic layer and serves as a protective "skin" for the "intentional unfinishedness" of the project.

项目名称：Žnjan-Pazdigrad Primary School / 地点：Pazdigradska 1, 21000 Split, Croatia / 建筑师：x3m-Mirela Bošnjak, Mirko Buvinić, Maja Furlan Zimmermann / 结构工程师：Eugen Gajšak, Marija Šarac, Marijan Bračun, Ivan Dolovčak / 工程：Siniša Radić, Milan Bjedov, Branimir Cindori, Marin Blažetić / 电气：Vojislav Štrbac / 消防顾问：Ognjen Truta / 景观设计师：Ines Hrdalo,Vesna Hrga Martić / 建筑物理：Mateo Biluš / 建筑技术：Zoran Divjak, Branislav Trifunović / 建造顾问：Teo Cvitanović / 土木、岩土、勘测：Alojzije Car / 项目管理：Teo Vojković, Lada Peranović, Mislav Olujić, Bruno Tudor, Vlatko Šokota / 字体设计：Nikola Đurek / 客户：City of Split / 用途：education, school / 用地面积：11,600m² / 建筑面积：5,225m² / 总建筑面积：7,830m² / 造价：900 EUR/m² / 设计时间：2009
施工时间：2015—2017 / 摄影师：©Bosnić + Dorotić (courtesy of the architect) (except as noted)

TTC 槟椥精英幼儿园

TTC Elite Ben Tre Kindergarten

KIENTRUC O

KIENTRUC O建筑事务所一向引以为豪的是他们对环境的强烈关注。在建筑设计过程中，这一点非常重要。因此，反过来说，一个已建成的项目将会对它所处的建筑环境做出有影响力的贡献。TTC槟椥精英幼儿园所营造的幽静氛围就是这样的积极贡献，它实现了将人、自然和槟椥地区文化汇聚一身的建筑潜力。

越南的Bao半岛气候温和，位于美丽的九龙江冲积平原之上。建筑师利用这一景观的壮丽之美，根据越南的传统建筑原则进行设计，这其中通常包括一个开放式平面设计，在其剖面上用水平面清晰地描绘出来，使建筑融入到更大的生态环境中。

设计这样的空间需要考虑到以儿童为中心的校园应该是个什么样子，以及怎样才能实现等问题。在这个环境中，孩子们应该能够轻松地拥有表达自我的自由，在一个有趣的环境下进行探索和学习，以使他们发挥出最好的自己。

从表现形式上来说，幼儿园被划分为两个独立的建筑体量，一个以斜坡的形式存在，一个以位于山上的几何形建筑体量的形式存在，它们为体育和艺术活动提供了灵活的空间，并且可以观赏城市美景，一览无余。

山下的空间被用作教室，由一个大型多用途大厅连接，该大厅是幼儿园的重要组成部分。这里可以每天进行各种教育活动，也可以举办特殊场合的活动。在这个多功能的空间内，有一个大型的景观操场，它友好地向各种元素开放，创造出一个可以让父母和孩子全天互动的理想场所。

槟椥是九龙江三角洲的一个相对平坦的地区。在其周围环境中，山的形象为城市提供了一种清新的建筑美学。因为学校面向Truc Giang湖，校园通过向上倾斜的矩形建筑体量逐渐改变了城市现有的结构，在当地居民和在山上玩耍的孩子之间建立了生动的视觉联系。

山的存在是帮助学校建立建筑特征的基础。它被设想成一个能调动孩子们的感官并激发好奇心的地方，让他们好奇，期待着山那边有什么令人兴奋的事情在等着他们。

KIENTRUC O pride themselves on their strong regard for context. It is an important influencer in the process of building creation, such that, in return, a built project will make an impactful contribution to the built environment it inhabits. The peaceful atmosphere realized at TTC Elite Ben Tre Kindergarten is one such positive contribution; it realizes the potential for architecture to integrate people, nature, and the generous culture of Ben Tre.

Bao peninsula, Vietnam, is characterized by a temperate climate, and is situated in one of the alluvial plains of the beautiful Cuu Long River. Engaging with the sublime beauty

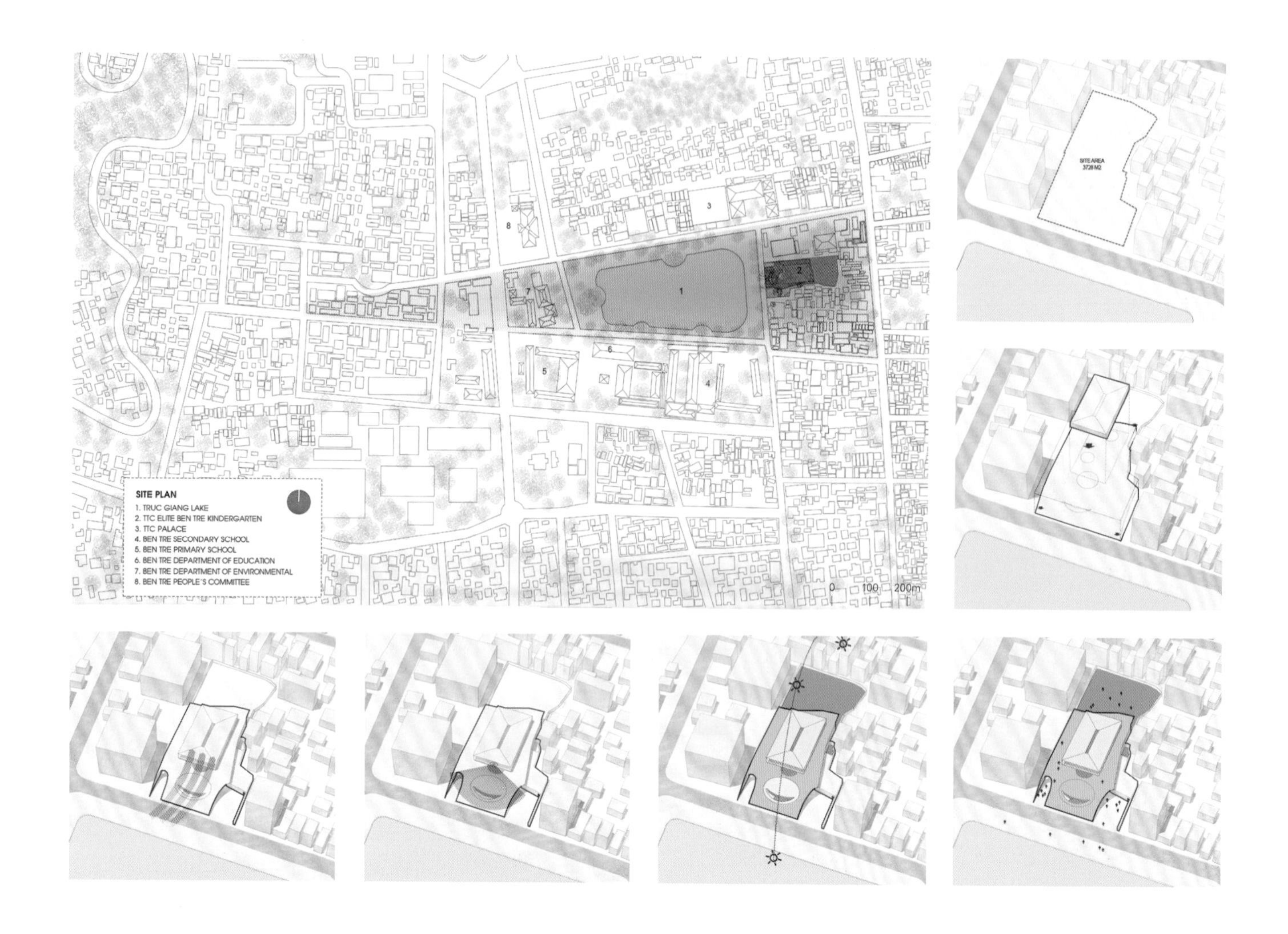

of this landscape, the architects implement principles from traditional Vietnamese architecture, which typically consists of an open plan, clearly delineated by horizontal planes in its section, allowing the architecture to infuse into a larger ecological context.

Designing such spaces involves a consideration of the implications of what, and how, a child-oriented educational campus should be like. Children should have the freedom to express themselves easily, explore and learn in a playful environment that brings out the best in them.

In a figurative sense, the kindergarten is compartmentalized into two separate masses, a sloping hill form and a geometrical mass situated above the hill – offering flexible spaces for physical and artistic activities with an uninterrupted view across the city.

The spaces below the hill serve as classrooms linked by a large multi-purpose hall, which is a central component of the kindergarten. It allows for various educational activities to take place on a daily basis, and on special occasions. Within this versatile space is a large landscaped playground that pleasantly opens up to the elements, creating an ideal place for parents and kids to interact with each other throughout the day.

Ben Tre is a relatively flat region within the Cuu Long delta. In its immediate context, the image of the hill offers a refreshing architectural aesthetic for the city. As it faces the Truc Giang lake, the campus gradually alters the city's existing urban fabric by sloping upward toward the rectangular mass, establishing a lively visual connection between the local residents and the children playing on the hill.

The presence of the hill is the corner-stone that helps found the architectural identity of the school. It is conceived as a place that juggles the senses and tickles the curiosity of each child, leaving them wondering and anticipating what excitements await them beyond the hill.

DAM 11/17
DAM 2017.

TUYỂN SINH

项目名称：TTC Elite Ben Tre Kindergarten / 地点：Ben Tre City, Vietnam / 建筑师：KIENTRUC O / 总建筑师：Đàm Vũ
设计团队：An-Ni Lê, Việt Nguyễn, Phương Đoàn, Duy Tăng, Dân Hồ, Tài Nguyễn, Giang Lê / 客户：TTC EDU
用途：kindergarten / 用地面积：3,728m² / 建筑面积：1,491m² / 竣工时间：2017 / 摄影师：©Hiroyuki Oki (courtesy of the architect)

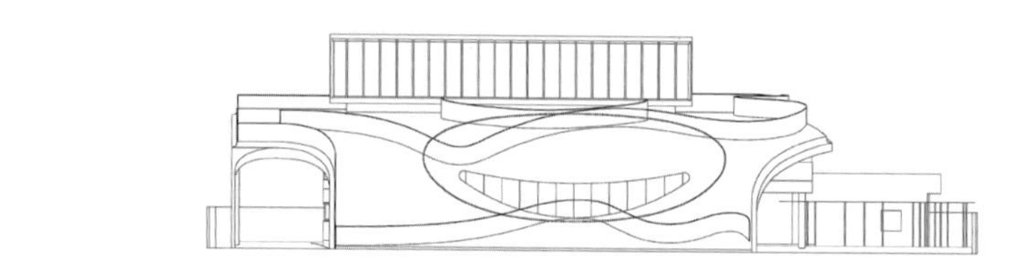

东立面 east elevation

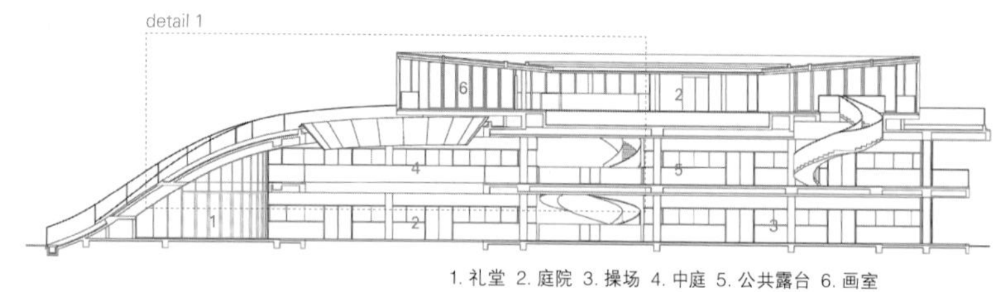

1. 礼堂 2. 庭院 3. 操场 4. 中庭 5. 公共露台 6. 画室
1. auditorium 2. courtyard 3. playground 4. atrium 5. communal terrace 6. painting room
A-A' 剖面图 section A-A'

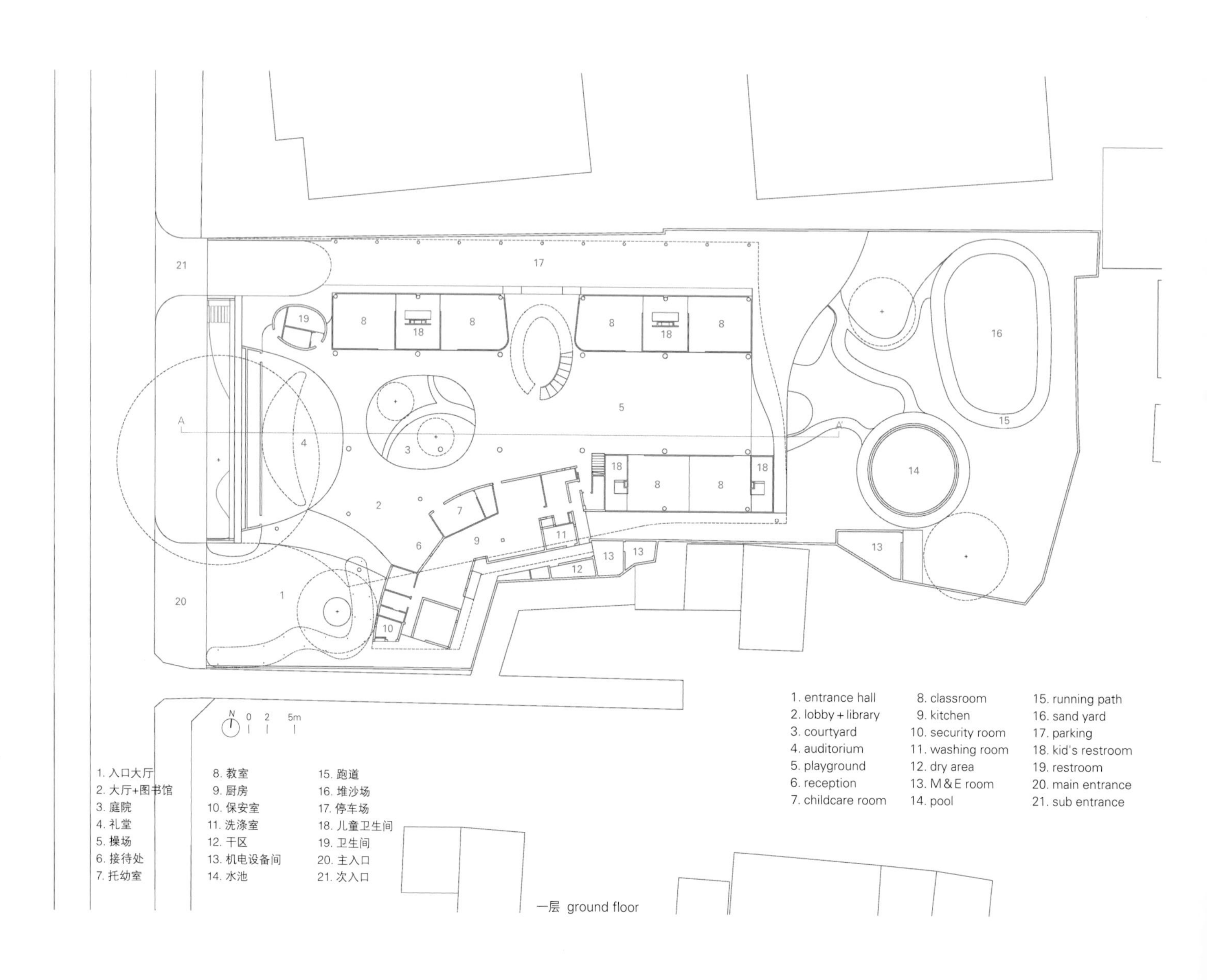

1. entrance hall
2. lobby + library
3. courtyard
4. auditorium
5. playground
6. reception
7. childcare room
8. classroom
9. kitchen
10. security room
11. washing room
12. dry area
13. M & E room
14. pool
15. running path
16. sand yard
17. parking
18. kid's restroom
19. restroom
20. main entrance
21. sub entrance

1. 入口大厅
2. 大厅+图书馆
3. 庭院
4. 礼堂
5. 操场
6. 接待处
7. 托幼室
8. 教室
9. 厨房
10. 保安室
11. 洗涤室
12. 干区
13. 机电设备间
14. 水池
15. 跑道
16. 堆沙场
17. 停车场
18. 儿童卫生间
19. 卫生间
20. 主入口
21. 次入口

一层 ground floor

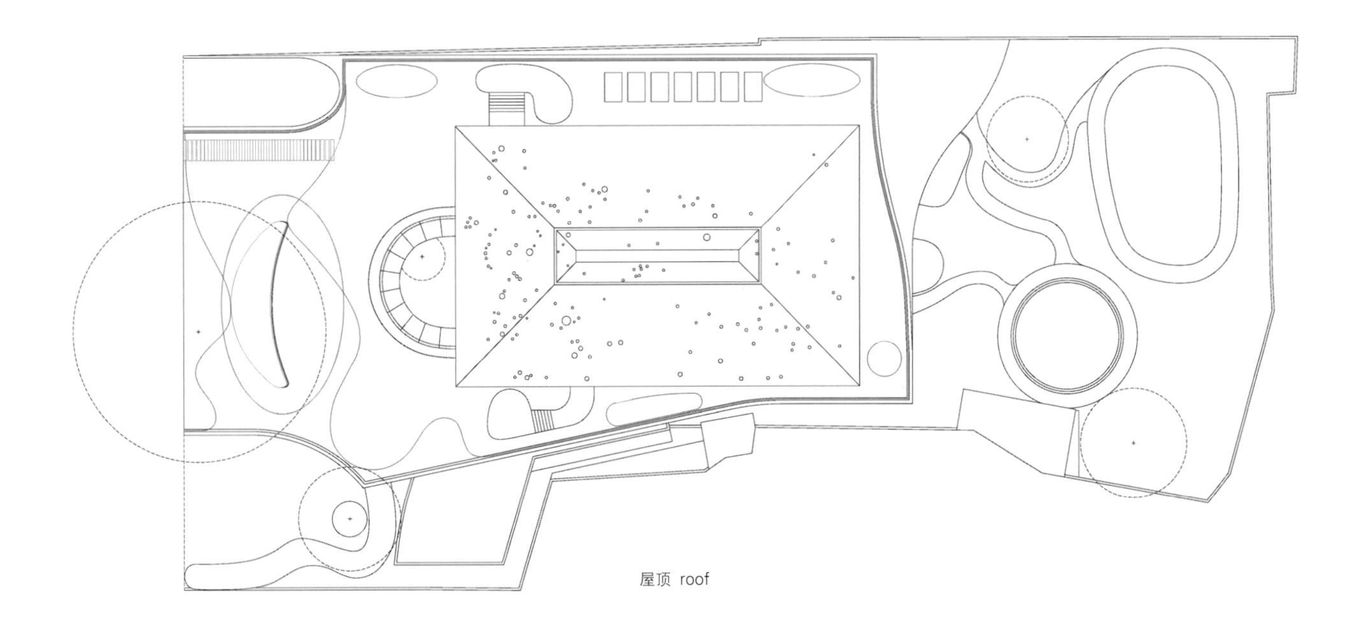

屋顶 roof

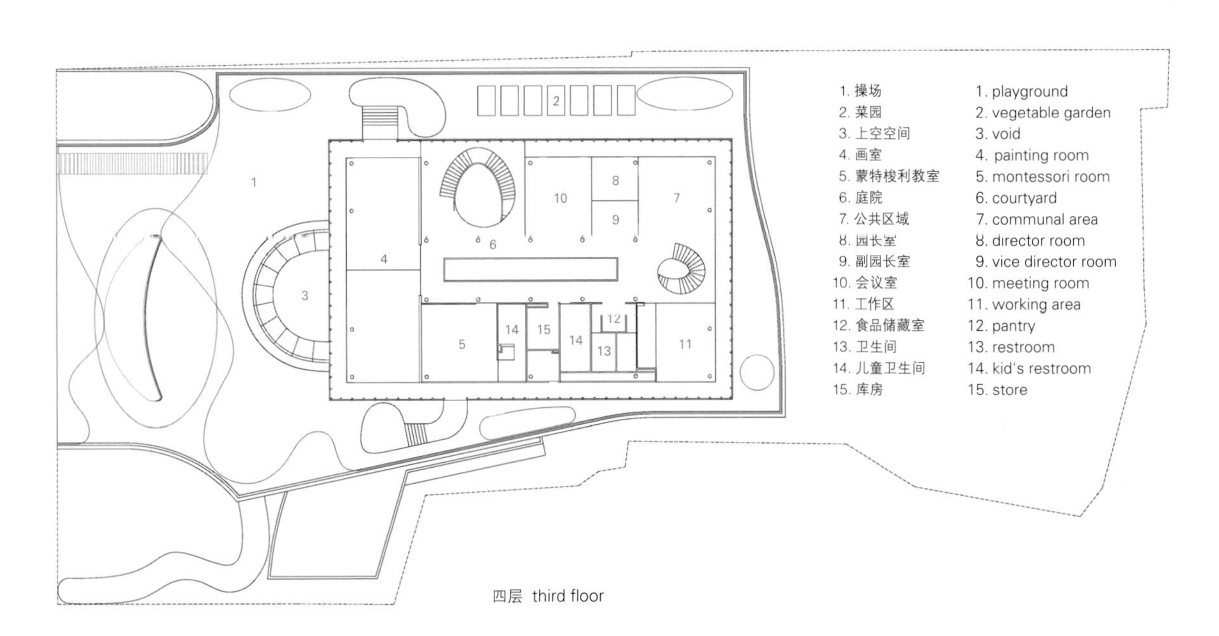

四层 third floor

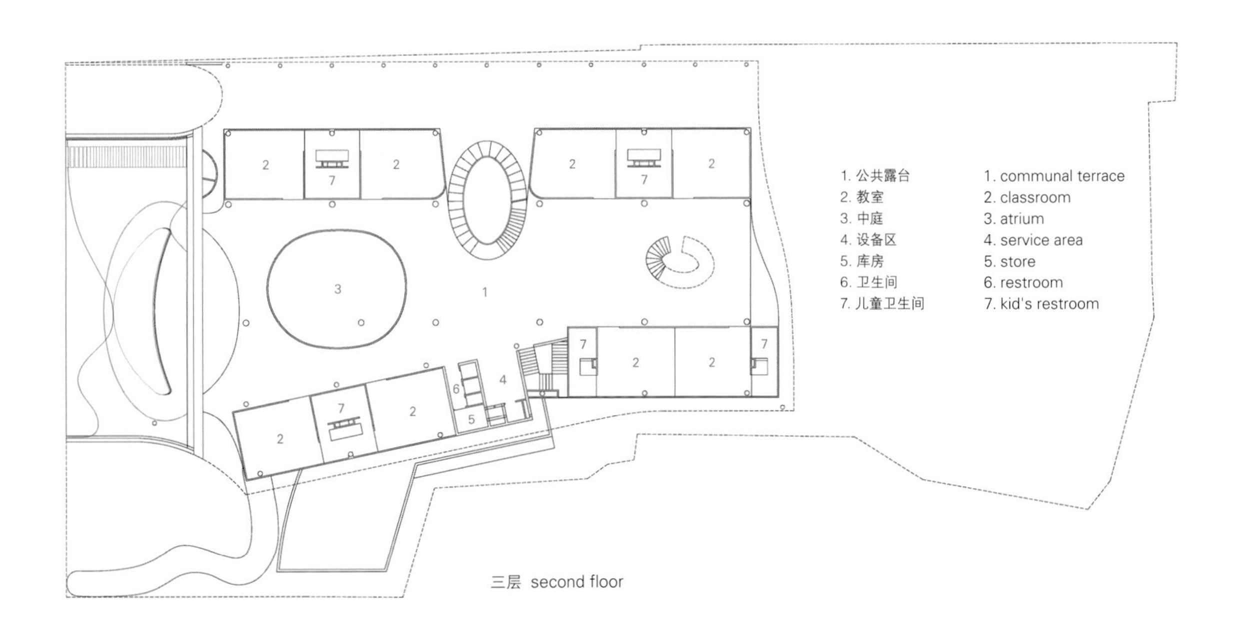

三层 second floor

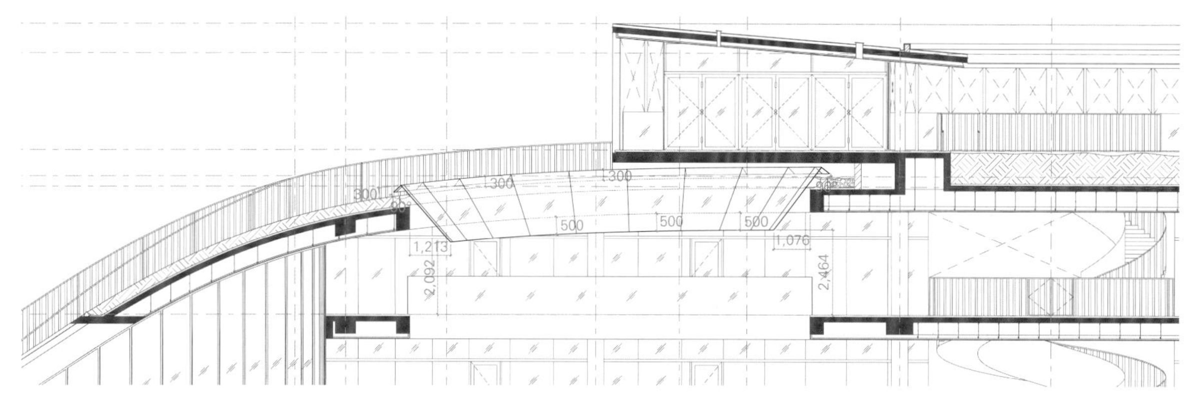

详图1——漏斗结构 detail 1_funnel

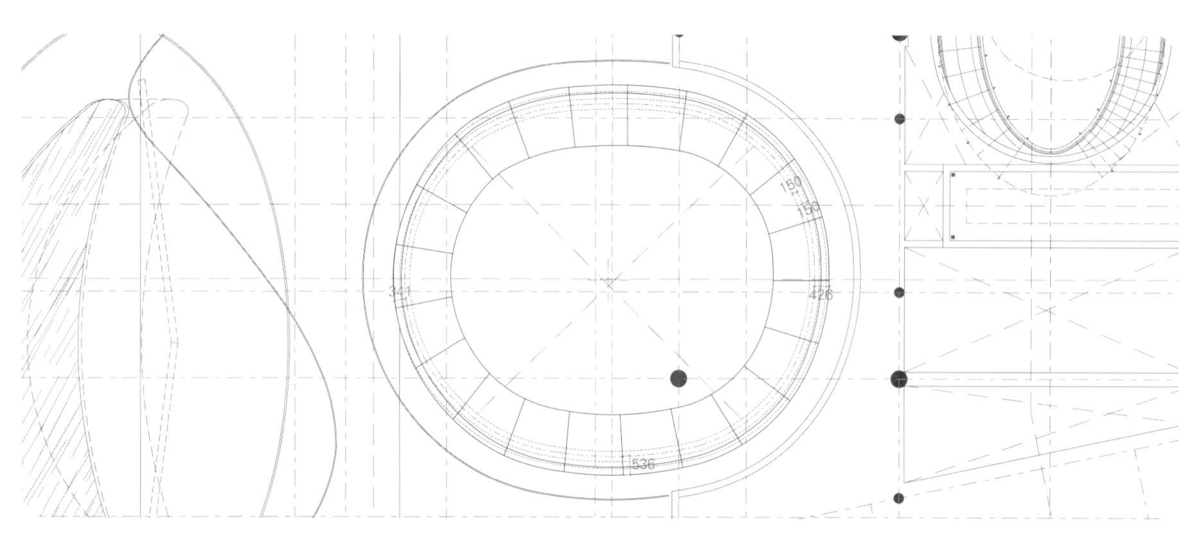

漏斗结构平面图 funnel plan

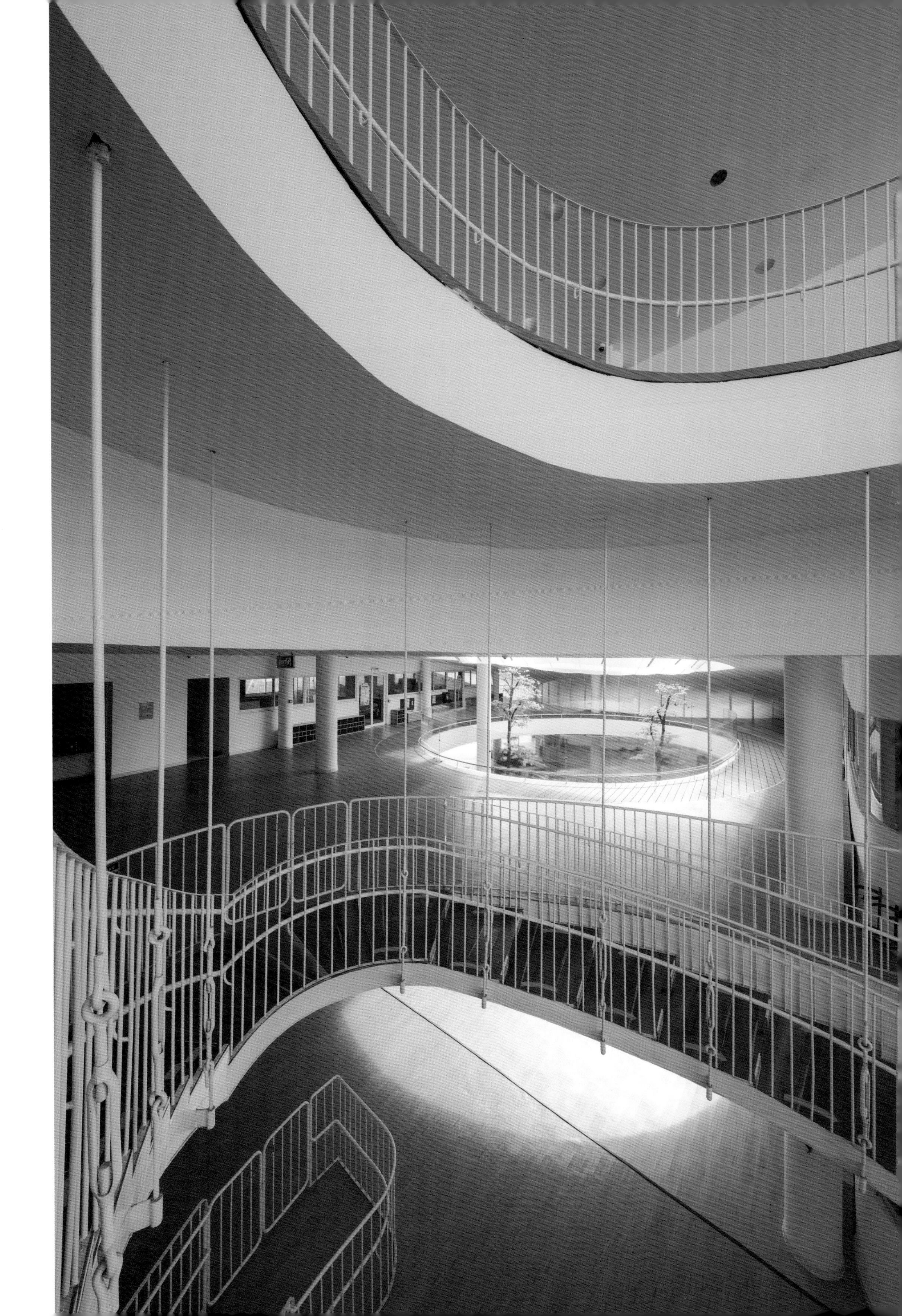

LÊN XUỐN
CẦU THAN

Nová Ruda 幼儿园

Kindergarten Nová Ruda

Petr Stolín Architekt

位于捷克共和国Vratislavice nad Nisou的新建成的Nová Ruda幼儿园，通过为儿童提供教育空间和休憩用地来满足城市日益增长的人口需求。

这所幼儿园建在利贝雷茨郊区的一个朝北的缓坡上。场地和其他方面的限制引发了一项关于空间试验的提议，该试验让孩子们看到了建筑设计的所有不同的形式。

建筑师希望孩子们通过幼儿园中各种特殊的结构来学习和感知建筑。孩子们能够逐渐发现整个幼儿园，并在进入大楼后明确自己的方位。

孩子们能够用一整天的时间发现这座建筑以及所有可能的进出的方法。教室不是普通的长方形房间。这是一个空间，通过大型平行窗户与室外在视觉上相连，通过中庭与其他楼层在视觉上相通。

孩子们所能进行的活动因身处建筑的不同楼层而不同：在安静的下部楼层是有床的睡眠区。上面的楼层是游戏室的所在。在这里，孩子们也有机会到外面的露台上，沿着建筑立面周围及后方的侧廊往前走。在那里，他们可以到达位于餐厅上方的主要室外游戏露台。实际上，整座建筑就像是一个等待探索的大丛林一般。

顶层的净高是最高的，可以作为教室使用。天窗为这里和地面层的餐厅带来了自然光线。该建筑的设计容纳了2个部门的25名儿童、供4名教育工作者使用的办公室、供1名托儿所经理和园丁使用的办公室，以及一个可容纳30名儿童和4名成人的餐厅。

由于其露台和庭院的设计，孩子们可以在户外阳光下玩耍。这些区域还配备了用于存放玩具和器具的室外储藏室。

幼儿园两个建筑主体之间的空间作为露台和公共空间使用。该区域连接着新的停车场，以及与建筑轴线对齐的外部木质地面。其余地方将用作绿地，并根据功能要求提供各种不同的游戏元素。

这所幼儿园被设计成一座砖砌建筑，在木格栅上安装了一层玻璃幕墙。整座建筑由一个钢结构包围，两条步行道环绕两座主体建筑，而另一层梯形玻璃纤维将主体建筑从街道处隐藏起来。

这种通透的设计概念增强了整体的建筑体量，并创造了一种安全感。外立面就如同一层柔软的面纱，包裹着幼儿园的内部世界。外壳将两个隔间纤薄的体量融合在一起，它们在后面通过公共区域和功能区域相连。这种联系使内中庭成为供儿童和成人使用的安全而又舒适的空间。

The new Nová Ruda Kindergarten in Vratislavice nad Nisou, Czech Republic, responds to the city's growing population needs by providing educational spaces and a leisure area for children.

The kindergarten was built on a north facing, gently sloping, site on the outskirts of Liberec. The limits of the site and other constraints led to a proposal for a spatial experiment that taught children to see architecture in all of its different forms.

The architects wanted the children to learn and perceive architecture through the various special configurations of a kindergarten. Children are able to discover the whole kindergarten gradually and orient themselves after entering the building.

All day long, the children can discover the building and all the possible ways to move in and around it. A classroom is not an ordinary rectangular room. It is a space that is visually connected to the exterior by large parallel windows, and to other floors through atriums.

项目名称： Kindergarten Nová Ruda / **地点：** Donská street, Nová Ruda, Vratislavice nad Nisou, Liberec, Czech Republic / **建筑师：** Petr Stolín Architekt s.r.o. / **主持建筑师：** Petr Stolín, Alena Mičeková / **客户：** City District Vratislavice nad Nisou / **建筑面积：** 425m² / **总建筑面积：** 915m² / **室外面积：** 770m² / **造价：** 1.5 mil euro / **竣工时间：** 2018
摄影师： ©Alexandra Timpau_Alex shoots buildings (courtesy of the architect)

MATEŘSKÁ ŠKOLA NOVÁ RUDA
N
0
20
50m

The children's activities vary across the floors of the building: in the quiet lower part is a sleeping area with beds. The floor above houses the playroom. Here, too, the children have the opportunity to go outside onto a terrace and along side galleries which continue around and behind the facade of the building. There they can reach the main outdoor play terrace situated above the dining room. The whole building is, in effect, one big jungle to explore.
The top floor has the highest clear height and serves as a classroom. Skylights bring natural light here and also to the dining room on the ground floor. The building is designed to accommodate 2 departments of 25 children, an office for 4 educators, an office for the nursery manager and farmer and a dining room for 30 children and 4 adults.
The facility, thanks to its terraces and courtyard, allows for outdoor play in the sun. These areas are also equipped with outdoor storage for toys and utensils.
The space that lies in between the two main bodies of the kindergarten functions as a terrace and public space. This area is connected to the new parking lot, and exterior wooden paving that is aligned to the axis of the building. The remaining terrain will be used as green space with various play elements required by the program.
The kindergarten is designed as a brick building with a fiberglass facade on a wooden grid. The whole building is embraced by a steel structure with two walking paths around the two main sections hidden from the street by another layer of trapezoidal fiberglass.
This concept of transparency enhances the whole mass and creates a sense of security. The outer facade is a soft veil that embraces the inner world of the kindergarten. The outer shell blends the two slim volumes of the two compartments, which are connected at the back by common and functional areas. This connection turns the inner atrium into a sheltered and pleasant space for children and adults alike.

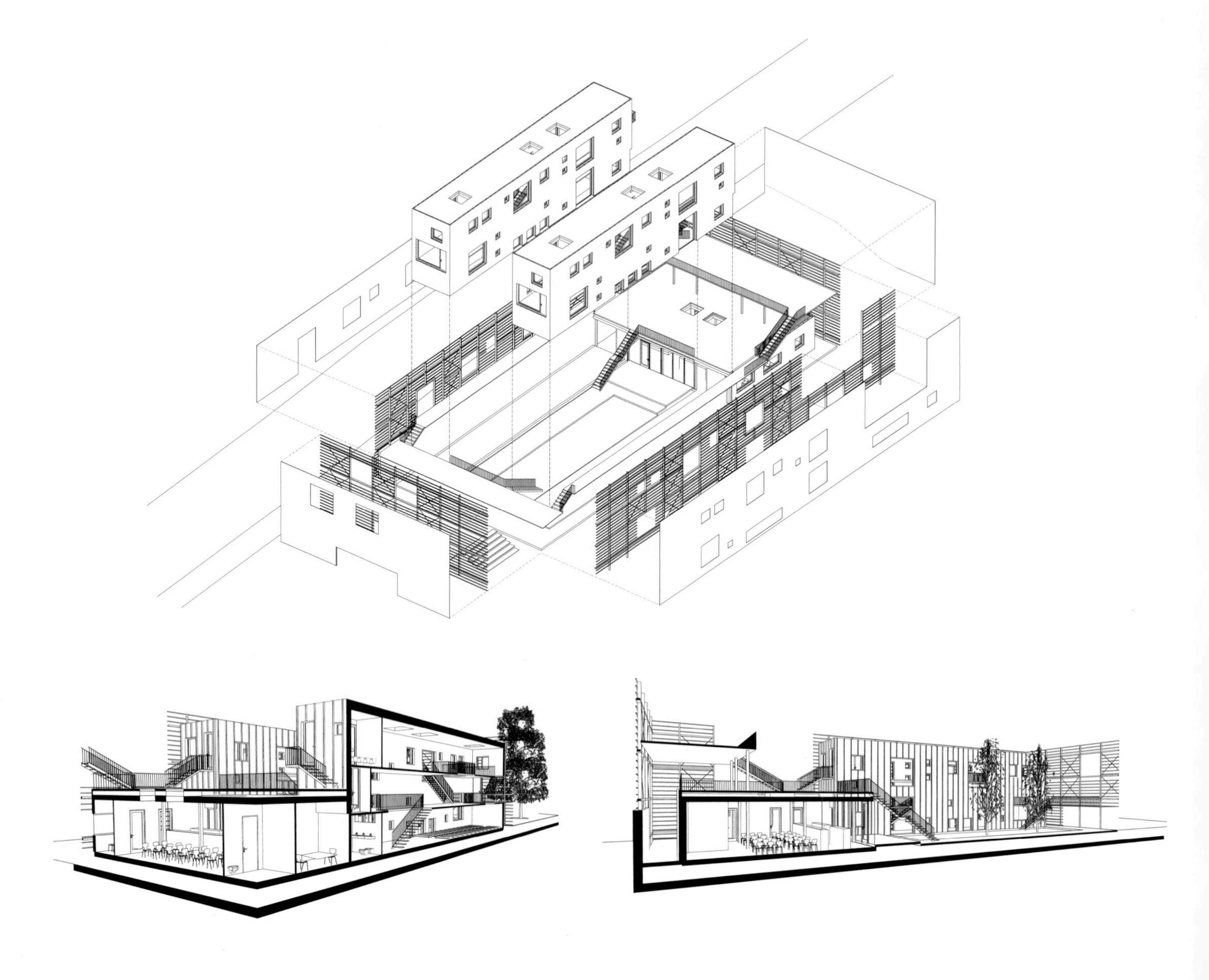

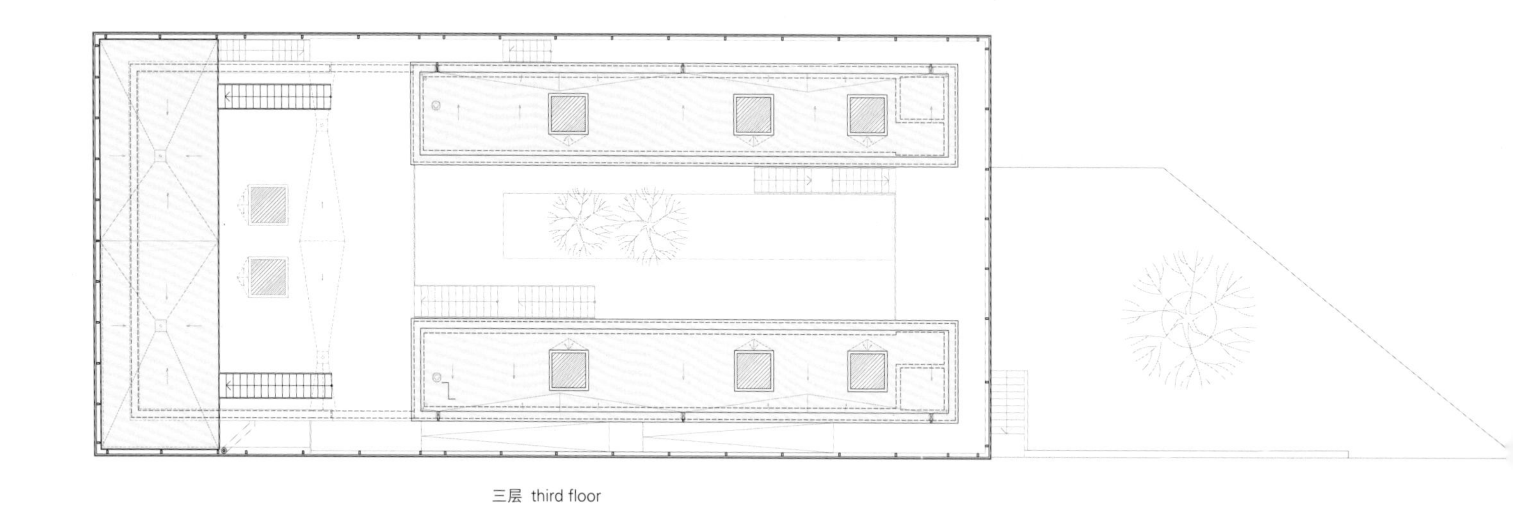

三层 third floor

二层 second floor

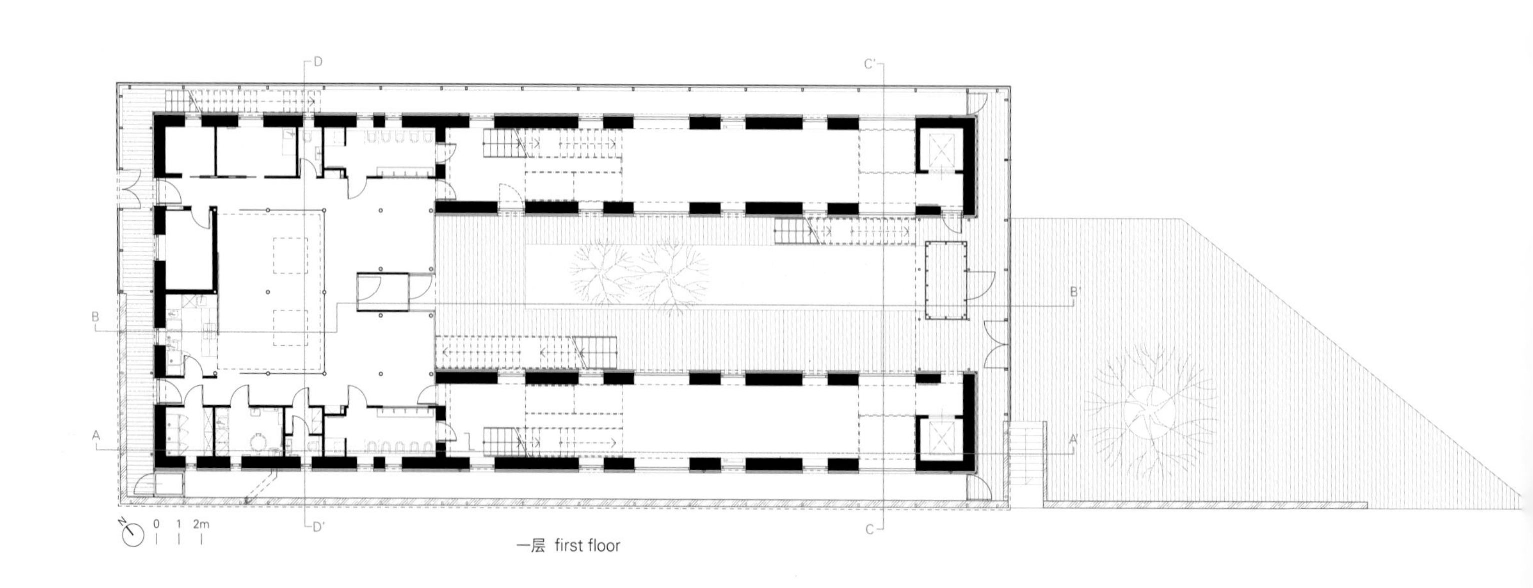

一层 first floor

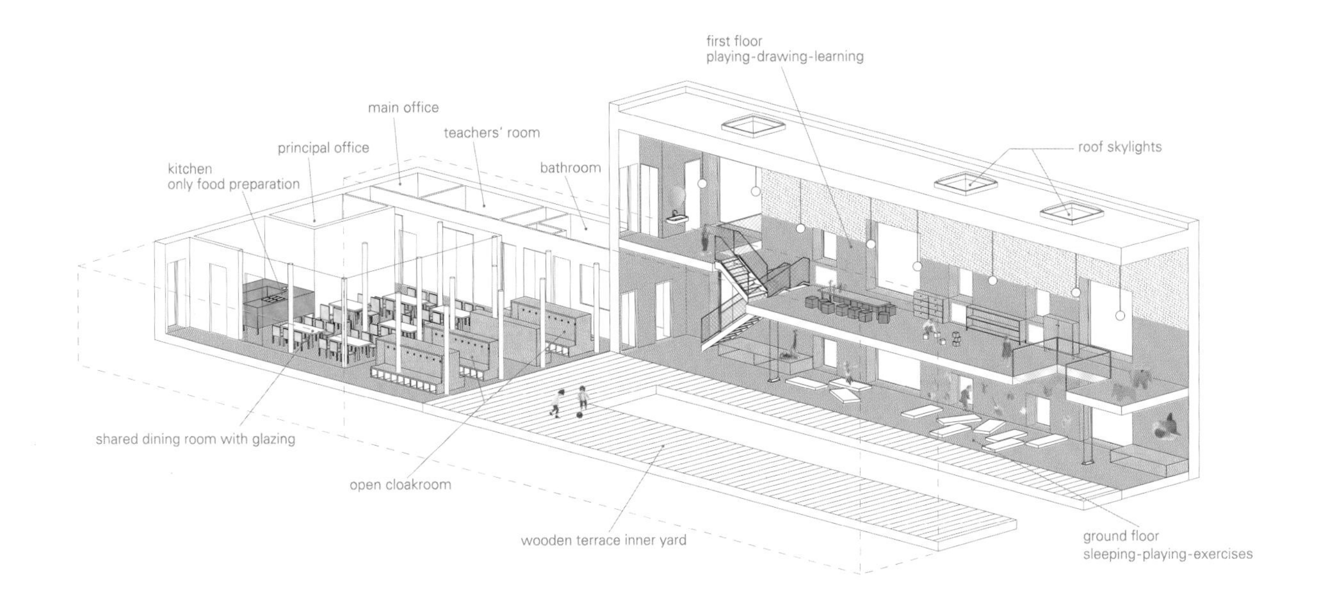

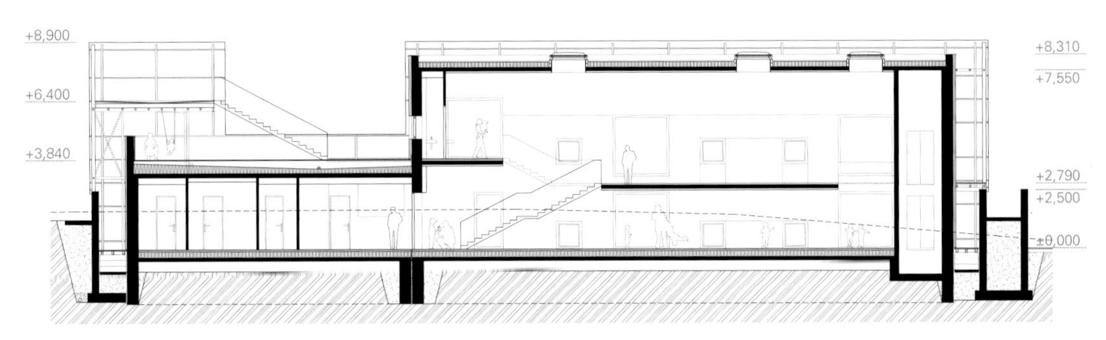

A-A' 剖面图 section A-A'

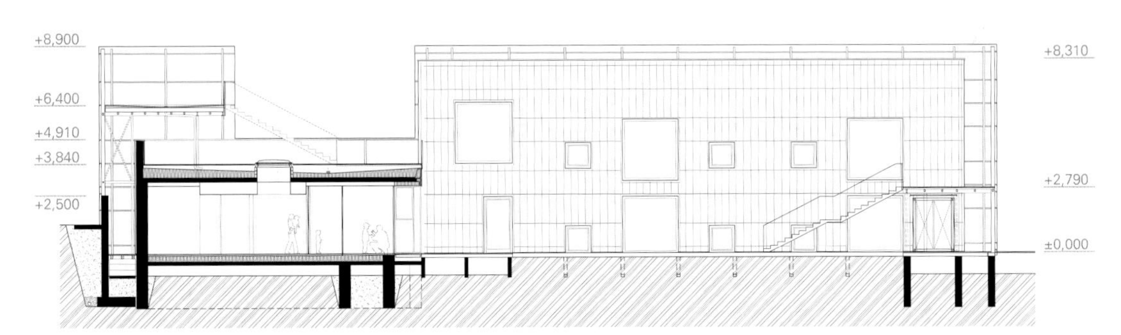

B-B' 剖面图 section B-B'

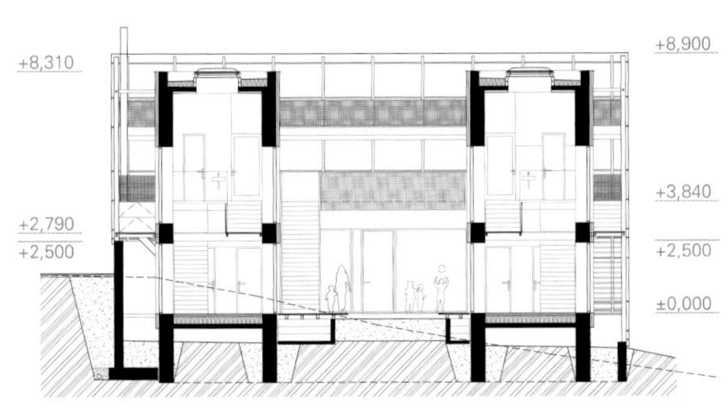

C-C' 剖面图 section C-C'

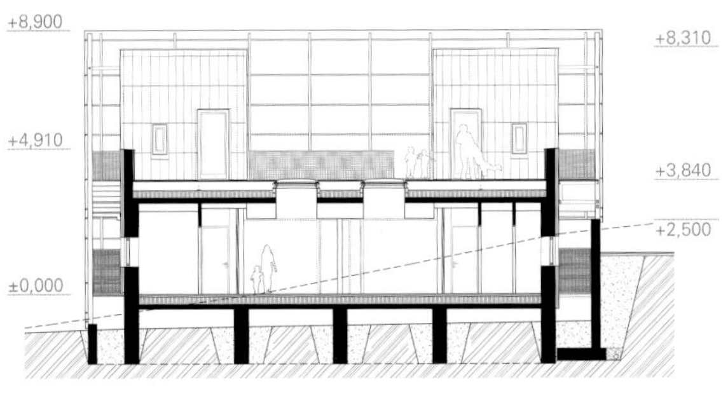

D-D' 剖面图 section D-D'

P208 **KIENTRUC O**

Was co-founded by dynamic duo of principal architect, ĐÀM VU[picture-above] and industrial designer LÊ AN-NI in Sài Gòn, Vietnam. ĐÀM VU graduated from the University of Architecture, Saigon in 2001 and worked for 4 years at the T.A.D. Founded One Architecture Studio in 2007 and reformed it to KIENTRUC O in 2013. Has been teaching at the University of Architecture Ho Chi Minh City since 2015. LÊ AN-NI studied architecture at the University of Architecture, Saigon and Industrial Design at the Faculty of Engineering, LTH (Lund University). Has been teaching Industrial Design at the University of Architecture Ho Chi Minh City since 2014.

P156 **Anna Roos**

Studied architecture at the UCT (University of Cape Town) and holds a postgraduate degree from the Bartlett School of Architecture, UCL, London. Moving to Switzerland in 2000, she worked as an architect, designing buildings in Switzerland, South Africa, Australia, and Scotland. As a freelance architectural journalist since 2007, besides C3, she also writes for A10, Ensuite Kultur Magazin, Monocle, and Swisspearl architecture magazine. Her first book, *Swiss Sensibility: The Culture of Architecture in Switzerland* (2017), published by Birkhauser Verlag.

P222 **Petr Stolín Architekt**

Petr Stolín[left] studied at the Faculty of Architecture of Brno University of Technology, Czech from 1978 to 1983. Worked in atelier of Pavel Švancer at Stavoprojekt Liberec and at SIAL under Karel Hubáček. Has been running his independent architectural practice since 1993. Often works with his brother, a sculptor and conceptual artist Jan Stolín. Currently he runs atelier at Faculty of Arts and Architecture of Technical University of Liberec. Alena Mičekova[right] has participated on some projects since 2003 as a part of her student practice and graduated from the Faculty of Arts and Architecture of Technical University in Liberec. Has been working in the atelier since 2007.

P10 **Michèle Woodger**

Is a writer based in London. Studied as an undergraduate at the University of Cambridge and as a postgraduate at UCL. Was previously the editor of a market-leading construction publication and website, and has worked in architecture publishing for ten years. Has recently been awarded research funding from the RIBA Gordon Rickett's Trust and the Society of Architectural Historians of Great Britain (SAHGB) to study public lettering in Lon-

two decades is characterised by its ingenuity, inventiveness and originality. Has been appointed a Commander of the Order of the British Empire, a Royal Academician and in 2004, became the youngest Royal Designer for Industry. Group Leader, Stuart Wood joined Heatherwick Studio in 2002. Has led the UK Pavilion for the 2010 Shanghai World and the design phases of Google's two new 1 million+ sqft headquarters in California, now in construction. Project Leader, Laurence Dudeney is currently leading two projects in New York for the Related Properties; 515 West 18th Street, a two-tower residential development that crosses both sides of the High Line, and Vessel, a new landmark for Hudson Yards.

Thomas Heatherwick

©Earl Wan

P162 **JJW Arkitekter**

Is a Danish architecture studio established in 1986 by the architects Anders Holst Jensen, Peter Henning Jørgensen and Kaj Frederik Wohlfeldt with the name of Jensen+Jorgensen+Wohlfeldt. Changed its name to JJW Arkitekter in 2004. Today the practice employs 75 architects, constructing architects, landscape and interior architects. Is specialized within sustainability and DGNB-certification, user involvement and process management. Works with education, housing, healthcare, and business and aims to design long-lasting buildings that doesn't burden future generations. (Lars Andersen, Katja Viltoft, Anders Holst Jensen, David Ploug and Ole Hornbek, from left, in the picture)

P178 ANMA

Since its foundation in 2001, ANMA's partners, Nicolas Michelin[left], Michel Delplace[right] and Cyril Trétout[center], have been developing innovative studies and productions in the fields of architecture, urban planning, and landscaping. Forms a network of designers, researchers, and artists as well as spaces for working, exchanging ideas and experimenting. Based in Paris, Bordeaux and Beijing, they produce ultra-contextual urban projects marked by constructive rigour and an unflagging determination to use natural energies. Their approach does not depend on style or technique but mirrors an attitude that each project is customized to reflect the site and its users. ANMA is a partner in research programs at the University of Montreal and the University of Shenyang in China and with Eurhonet, an international group of social housing operators.

P50 Davide Pisu

Is an Italian architect, PhD candidate at the University of Cagliari and visiting PhD candidate at the University of Hertfordshire. He currently is leading research on the relations between norms and architectural form. His research interests are centred on architectural design and theory, with a focus on the architecture of information and knowledge. His architectural works include housing and public spaces, as well as an ongoing collaboration with C+C04 Studio.

©Tuomas Uusheimo

P56 ALA Architects

Is a Helsinki based architectural office founded in 2005. The office specializes in demanding public and cultural buildings, station design, and unique renovation projects. ALA is run by three partners: Juho Grönholm[center], Antti Nousjoki[left] and Samuli Woolston[right], and in addition to them employs a team of 51 architects, interior designers, students and staff members. ALA is best known for their Kilden Performing Arts Centre in Kristiansand, Norway, completed in 2011. Their ongoing projects include the expansion of Helsinki Airport, three metro stations in Espoo, Finland and a hotel in Tampere, Finland.

P110 Herzog & de Meuron

Was established in Basel, 1978. Is led by Jacques Herzog and Pierre de Meuron with Senior Partners Christine Binswanger, Ascan Mergenthaler, Stefan Marbach, Esther Zumsteg, and Jason Frantzen. An international team of five Partners, about 40 Associates and 400 collaborators works on projects across Europe, the Americas and Asia. Has been awarded numerous prizes including the Pritzker Architecture Prize in 2001, the RIBA Royal Gold Medal and the Praemium Imperiale (Japan) in 2007. In 2014, received the Mies Crown Hall Americas Prize and was awarded the title of doctor honoris causa in 2018 by the Technical University of Munich.

P128 MX_SI

After ten years collaboration since 2005, MX_SI(Mendoza Partida + BAX studio) has decided to open two independent studios.
Mendoza Partida is an international architecture practice based in Barcelona. Experimentation, investigation and innovative solutions based on common sense and consistency give their architecture a precise balance of geometry, matter and emotion.
BAX studio is run by Boris Bežan and Mónica Juvera who developed international projects from Spain, Slovenia, Finland, Norway to Mexico city. Is specialized in creating unique and coherent architecture that spans from social, landscape, structure to sustainable design. Won several international awards including a nomination for the Mies van der Rohe Award in 2017.

P104 Gihan Karunaratne

Graduated from Royal College of Arts and Bartlett School of Architecture. Has taught and lectured in architecture and urban design in UK, Sri Lanka and China. Writes and researches extensively on art, architecture and urban design. Was the Director of architecture for Colombo Art Biennale 2016. Exhibited in Colombo Art Biennale in 2014, Rotterdam Architecture Biennale in 2009 and the Royal Academy Summer Exhibition. Is a recipient of The Bovis and Architect Journal Award for architecture and a Fellow of Royal Society of Arts (RSA) for Architecture, Design and Education in 2012.

P82 Snøhetta

Kick-started its career in 1989 with the competition winning entry for the new library of Alexandria, Egypt. This was later followed by the commission for the Norwegian National Opera in Oslo, and the National September 11 Memorial Museum Pavilion at the World Trade Center in New York City, among many others. Since its inception, the practice has maintained its original trans-disciplinary approach, integrating architectural, landscape and interior design in all of its projects.
Recently completed works include the redesign of the public space in Times Square, the Lascaux IV Caves Museum in Montignac, France, and the expansion to the San Francisco Museum of Modern Art.
Received the World Architecture Award, Aga Kahn Prize for Architecture, the Mies van der Rohe EU Architecture Prize, the European Prize for Urban Public Space, and The International Architecture Award.

P82 DIALOG

Is a multi-disciplinary design firm with studios across Calgary, San Francisco, Vancouver, Edmonton and Toronto. The firm's work includes designing for urban vibrancy, health and wellness, transportation, education, arts and culture, residential, retail, and commercial, as well as mixed-use solutions. Is passionate about design. They believe it can, and should, meaningfully improve the wellbeing of our communities and the environment we all share.

P144 BAAS Arquitectura

Founder and director, Jordi Badia was born in Barcelona, 1961 and graduated in architecture from the ETSAB in 1989. After graduation, he has had a professional partnership with Tonet Sunyer until 1993 and founded BAAS in 1994. Has been a Senior Lecturer at ELISAVA-UPF (1989 – 2001), Associate lecturer at ETSAV (1994 – 2001), Lecturer at ESARQ-UIC (2009 – 2013) and Associate lecturer at his alma mater (2001 ~). Along with Félix Arranz, he was the curator of the Catalan and Balearic pavilion at the 13th Venice Architecture Biennale 2012. Has been a Member of the Territorial Heritage Commission for the city of Barcelona since 2010. Currently 3 team leaders, Àlex Clarà, Alba Azuara, Jero Gutiérrez are guiding the firm with him.

P190 x3m

Is a Croatian Zagreb-based architecture studio established in 2003 by architects Mirela Bošnjak(1973)[left], Mirko Buvinić(1971)[right] and Maja Furlan Zimmermann(1973)[center]. They graduated from the Faculty of Architecture of the University of Zagreb in the late 1990s, and have received various awards and recognitions both for works built and competition projects. Their office is a small practice where the three partners are also the only employees. Nevertheless, they design and manage all their projects fully in-house. The Žnjan – Pazdigrad Primary School won several national architecture awards and was nominated for the 2019 EU Mies Award.

©Geordie Wood

P36 Diller Scofidio+Renfro

Was founded in 1981 at New York. Is led by four partners of Elizabeth Diller[picture-above], Ricardo Scofidio, Charles Renfro and Benjamin Gilmartin. Completed adaptive reuse of High Line and transformation of Lincoln Center for the Performing Arts and The Shed, the first multi-arts center and the renovation and expansion of the Museum of Modern Art MoMA. Was also selected to design the Centre for Music, located at the heart of London's Culture Mile. Has been distinguished with the Time Magazine's "100 Most Influential" list, the Smithsonian Institution's 2005 National Design Award, the Medal of Honor and the President's Award from AIA New York, and Wall Street Journal Magazine's 2017 Architecture Innovator of the Year Award. Ricardo Scofidio and Elizabeth Diller are fellows of the American Academy of Arts and Letters and the American Academy of Arts and Sciences, and are International Fellows at the RIBA.

图书在版编目(CIP)数据

颂扬多元化 ：当今学校建筑 ：汉英对照 / （英）安娜·鲁斯等编 ；曹麟，吴美萱译. — 大连 ：大连理工大学出版社，2020.5

（建筑立场系列丛书）

ISBN 978-7-5685-2523-7

Ⅰ. ①颂… Ⅱ. ①安… ②曹… ③吴… Ⅲ. ①教育建筑－建筑设计－汉、英 Ⅳ. ①TU244

中国版本图书馆CIP数据核字(2020)第063869号

出版发行：大连理工大学出版社
（地址：大连市软件园路80号　邮编：116023）
印　　刷：上海锦良印刷厂有限公司
幅面尺寸：225mm×300mm
印　　张：14.75
出版时间：2020年5月第1版
印刷时间：2020年5月第1次印刷
出 版 人：金英伟
统　　筹：房　磊
责任编辑：张昕焱
封面设计：王志峰
责任校对：杨　丹
书　　号：978-7-5685-2523-7
定　　价：298.00元

发　行：0411-84708842
传　真：0411-84701466
E-mail：12282980@qq.com
URL：http://dutp.dlut.edu.cn

本书如有印装质量问题，请与我社发行部联系更换。

墙体设计
ISBN：978-7-5611-6353-5
定价：150.00元

新公共空间与私人住宅
ISBN：978-7-5611-6354-2
定价：150.00元

住宅设计
ISBN：978-7-5611-6352-8
定价：150.00元

文化与公共建筑
ISBN：978-7-5611-6746-5
定价：160.00元

城市扩建的四种手法
ISBN：978-7-5611-6776-2
定价：180.00元

复杂性与装饰风格的回归
ISBN：978-7-5611-6828-8
定价：180.00元

内在丰富性建筑
ISBN：978-7-5611-7444-9
定价：228.00元

建筑谱系传承
ISBN：978-7-5611-7461-6
定价：228.00元

伴绿而生的建筑
ISBN：978-7-5611-7548-4
定价：228.00元

微工作·微空间
ISBN：978-7-5611-8255-0
定价：228.00元

居住的流变
ISBN：978-7-5611-8328-1
定价：228.00元

本土现代化
ISBN：978-7-5611-8380-9
定价：228.00元

都市与社区
ISBN：978-7-5611-9365-5
定价：228.00元

木建筑再生
ISBN：978-7-5611-9366-2
定价：228.00元

休闲小筑
ISBN：978-7-5611-9452-2
定价：228.00元

景观与建筑
ISBN：978-7-5611-9884-1
定价：228.00元

地域文脉与大学建筑
ISBN：978-7-5611-9885-8
定价：228.00元

办公室景观
ISBN：978-7-5685-0134-7
定价：228.00元